INTERNATIONAL UNION OF CRYSTALLOGRAPHY
TEXTS ON CRYSTALLOGRAPHY

INTERNATIONAL UNION OF CRYSTALLOGRAPHY
TEXTS ON CRYSTALLOGRAPHY

---

1. The Solid State: from Superconductors to Superalloys
A. Guinier and R. Jullien (Translated by W. J. Duffin)

The Solid State
From Superconductors to Superalloys

André Guinier
Emeritus Professor, Université Paris-Sud
Member of the Académie des Sciences

Rémi Jullien
Professor, Université Paris-Sud

Translated from the French (*La matière à l'état solide: des supraconducteurs aux superalliages*)
by
W.J. Duffin
Fellow of the University of Hull

INTERNATIONAL UNION OF CRYSTALLOGRAPHY
OXFORD UNIVERSITY PRESS
1989

Oxford University Press, Walton Street, Oxford OX2 6DP
Oxford New York Toronto
Delhi Bombay Calcutta Madras Karachi
Petaling Jaya Singapore Hong Kong Tokyo
Nairobi Dar es Salaam Cape Town
Melbourne Auckland
and associated companies in
Berlin Ibadan

Oxford is a trade mark of Oxford University Press

Published in the United States
by Oxford University Press, New York

British Library Cataloguing in Publication Data
Guinier, André
The solid state: from superconductors to superalloys.
1. Solid state physics
I. Title II. Jullien, Rémi III. Series
530,4'1

ISBN 0–19–855290–4
ISBN 0–19–855554–7 (pbk.)

Library of Congress Cataloging in Publication Data
Guinier, André.
[Matière à l'état solide. English]
The solid state: from superconductors to superalloys/André Guinier, Rémi Jullien; translated from the French by W.J. Duffin.
(International Union of Crystallography texts on crystallography; 1)
Translation of: La matière à l'état solide.
Bibliography. Includes index.
1. Solid state physics. I. Jullien, R. II. International Union of Crystallography. III. Title. IV. Series
QC176.G8513 1989 530.4'1—dc20 89–31661

ISBN 0–19–855290–4
ISBN 0–19–855554–7 (pbk.)

Typeset by Gecko Ltd, Bicester, Oxfordshire
Printed in Great Britain by
Butler & Tanner Ltd, Frome and London

Foreword

Sir Nevill Mott
Cavendish Laboratory, Cambridge

It is a pleasure and honour to write an introduction to this book by Professors A. Guinier and R. Jullien. The senior author is an old friend. I have lectured in France both before and after the war of 1939–45, and the first work of his that I remember is that on Preston–Guinier zones, a form of precipitate within an aluminium alloy that hardens it, and on the mechanism of which I made a theory (with Frank Nabarro) in 1939. During his long career Professor Guinier has proved himself a master in the wide field of solid state science, and this book deserves to have the same success as his earlier book on atomic physics for students (*The Structure of Matter*). It covers the solid state — perhaps the most rapidly growing branch of physical science and that with the closest links with modern technology. After chapters on specific heat, thermal expansion, and other subjects that were clarified in the early part of the century, it goes on to describe the properties of electrons in metals and semiconductors, the understanding of which is based on quantum mechanics and could hardly have developed without it. I am interested in the way the author handles this. He does not appear to assume that the student has already attended a course or read a book on quantum mechanics, as in the United Kingdom would, I believe, have usually been the case. Instead he introduces it from the beginning, but only just as much as is needed. I believe this is the right way, perhaps especially for the future engineers and technicians for whom the book will be especially valuable.

The last part of the book deals with diffusion, plasticity, dislocations, and Preston–Guinier zones. One feels here the contrast with semiconductors and solid state electronics, where theory went before practice. The metallurgists produced excellent heat-resisting alloys without the aid of the modern theory of dislocations, useful though that has now become.

In all advanced countries we need more and better educated scientists, engineers, and technologists, and the way we should teach them is under intensive discussion. In the United Kingdom, we have had for young people of ages between 11 and 18 the Nuffield Science project with its emphasis on learning by doing, and we debate both for school and university the relative advantages of a broad and a more narrow education. At Oxford the students study chemistry alone for four years, and one distinguished alumnus of their chemistry school thanked his good fortune in that this allowed him time to savour all other branches of human knowledge. In Cambridge we demand three experimental sciences and the timetable is rather too full for that to be easy. In the United States, too, a wider course is usual, at any rate for the first year.

I remember, in planning the Nuffield project for school science, a discussion of teaching for ages between 16 and 18; one group said that, if we did not include something on modern physics, relativity, and quantum mechanics, the most talented students would be bored and desert science for the humanities; others on our committee, who on the whole prevailed, said 'leave it to the universities; the duty of the schools is to give a sound understanding of classical science'. How and when to introduce students to modernity is a continuing problem. Perhaps Professor Guinier has the answer, in introducing modern theories as and when needed for practical problems. My first impression of the book was that it would be for students at the level of the second year of a British university. But perhaps, with the author's method of putting the mathematics into 'boxed' sections for a second reading and introducing quantum mechanics as and when needed to understand devices that every schoolboy handles, it might be used earlier.

Preface

There are so many good texts in the world dealing with the physics of solids at various levels that the authors of a new one in the same field need to justify their production of another. Our objective in writing the present volume is twofold.

In the first place, it is intended as a sequel to *The structure of matter*,[†] written in the same spirit and addressed in part to the same readership: non-specialist scientists and physics teachers concerned with the final years in secondary schools and the earlier years at universities, polytechnics, and colleges.

Our second aim is to provide material for readers who are engaged, or wish to engage, in a detailed study of one of the scientific or technical fields dealt with in the text. For such readers, our book will be an introduction to more advanced volumes, such as *Introduction to solid state physics* (Kittel, 1986). In order to tackle these, however, a level of understanding is required for which the almost completely qualitative approach adopted in the majority of the text is inadequate. For that reason, we have interspersed 'boxed' sections throughout the book to provide mathematical justification for some of the results that are merely stated in the main text, and to give a simple description of some useful calculations in solid state physics.

Our concern throughout has been to relate the macroscopic properties of solids (usually crystalline) to models of their atomic structure. This is not always possible: hence the very restricted choice of the phenomena included in the book and the omission of many details. We were keen to deal with mechanical properties alongside thermal, electrical, and magnetic properties, unlike the often separate treatment of the two groups of topics according to whether the book concerned has a 'metallurgical' or 'solid state physics' slant. In fact, such a distinction largely reflects the traditional compartmentalized way in which the subject is taught.

It is often claimed today that provision for thorough scientific and technical training is essential for any nation. Taking the necessary steps to increase the numbers of young people going into advanced science courses is, of course, the job of politicians. But scientists have a responsibility too: if we are content to hang on to styles of teaching suitable only for a restricted public gifted in the field of abstract calculation, many of those

[†]Guinier, A (1984). *The structure of matter: from the blue sky to liquid crystals*, Edward Arnold, London. References to this book in the present text are given in the form (SM, p. 47).

who might otherwise have been attracted to the sciences will quickly be discouraged and go elsewhere.

If science is really to be opened up to a wider public, scientists themselves must make greater efforts to ease the passage through the barriers formed by the difficulties surrounding their territory. There are cliffs around our subjects that only expert climbers can scale: we must carve out a path accessible to those of more modest talents. Those who engage in both teaching and research have a special responsibility for making science more accessible to greater numbers. That is what we have attempted to do for our own particular field in this book.

There is, of course, nothing original in these pages. Our sources are too numerous to quote, but we should like to offer our very sincere thanks to those colleagues whose publications have helped us or who have contributed to improvements in the manuscript through discussion and comment: Gérard Fournet, Jacques Friedel, Hubert Gié, Jacques Joffrin, Charles Mazières, and Jean Philibert. We are also very grateful to Yves Kaminsky, who offered us some perceptive criticisms from the point of view of a teacher. We also wish to express our warm thanks to those who have so willingly provided documents for the illustrations: Maurice Cagnon, Georges Cizeron, Charles Donadille, Jacques de Fouquet, Bernard Hennion, Guy Henry, Paul Lacombe, Charles Pénel, Jean Philibert, Irena Puchalska, Yves Quéré, and Colette Servant.

Finally, we wish to express our deep gratitude to Sir Nevill Mott, who not only willingly contributed a foreword to this book but also provided material for it on a subject to which he has made such an important personal contribution, thus demonstrating that recent advances in science can be explained in simple terms.

A.G
April 1987 R.J.

Translator's note

The section on superconductivity in Chapter 2 of the French edition has been revised and up-dated by Professors Guinier and Jullien for this edition. The new material takes into account recent advances in the field up to the beginning of 1989.

Contents

Titles of boxed sections x

1 Thermal properties of solids 1
- The heat capacity of a solid 1
- Thermal expansion 43
- Thermal conductivity 46

2 Electrical properties of solids 56
- Dielectrics 57
- The electrical conductivity of solids 69
- The electronic conductivity of metals 73
- Semiconductors 101
- Superconductivity 126
- New conductors 134

3 Magnetic properties of solids 136
- The action of a magnetic field on a solid 136
- Diamagnetism and paramagnetism 141
- Ferromagnetism 147
- Other types of magnetic ordering in crystals 156
- The behaviour of ferro- and ferrimagnetics in a magnetic field 160
- Relationships between magnetic properties and the structure of matter 169

4 Mechanical properties of solids 171
- General survey 171
- Elasticity 173
- Plastic deformation 190
- Fracture 220
- Creep 229
- Fatigue 234
- Some comments on the mechanical properties of metals 239

5 Diffusion 242
- The diffusion coefficient 245
- Diffusion in real systems 256
- Some applications of diffusion in solids 260

Bibliography 264

Index 265

Titles of boxed sections within the text

1 The principle of the equipartition of energy 3
2 The Bose–Einstein distribution 10
3 Longitudinal vibrations in a chain of identical atoms connected by springs 15
4 Definition of the reciprocal lattice and construction of Brillouin zones 19
5 Spectral density of low frequency vibrational modes. The heat capacity in Debye's model 25
6 Linear chain of two different atoms 35
7 Calculation of the thermal conductivity of a gas 51
8 Density of states for an electron in a box 79
9 Calculation of the Fermi level and the mean energy of free electrons 82
10 The Fermi–Dirac distribution and the electronic heat capacity 85
11 Bragg reflection of electron waves 92
12 The variation of the resistivity of a pure metal with temperature 98
13 Calculation of the number of carriers in a semiconductor 110
14 Classical calculation of the magnetic moment of an orbiting electron 138
15 Calculation of the magnetization as a function of temperature and applied field 144
16 Calculation of spontaneous magnetization using the molecular field approximation 150
17 Geometry of the extended spring 174
18 Expressions for the compressibility and rigidity in terms of Young's modulus and Poisson's ratio 178
19 Theoretical calculation of the elastic limit 197
20 Calculation of the theoretical stress needed for the cleavage of rock salt 222
21 The diffusion equation: solution to the bar problem 250

1

Thermal properties of solids

The heat capacity of a solid and its coefficients of thermal expansion and thermal conductivity form a group of properties related to effects produced by a variation in temperature. This chapter deals with such properties and their interpretation from an atomic viewpoint, beginning with the most fundamental of them: the heat capacity.

The heat capacity of a solid

The temperature of a solid rises when it is supplied with heat, and the heat is given up when the temperature falls: the solid is therefore said to have a certain **heat capacity**, which will depend on the mass involved. We shall be concerned with the molar heat capacity of a substance at a temperature T, measured by finding the amount of heat needed to raise the temperature of a mole of the substance from T to $T+1$ (T is in kelvin, K). In a solid, the heat supplied goes almost entirely into increasing the internal energy of the substance and *the molar heat capacity C can thus be defined as the derivative of the molar internal energy U of the substance with respect to temperature T, i.e.* $C = dU/dT$. (The internal energy always increases with temperature, so that C is always positive).

In the case of solids, this definition is adequate as it stands and there is no need to specify whether the temperature variation occurs at constant pressure or constant volume since the thermal expansion is extremely small. We recall that it is different with gases, e.g. for a monatomic gas:

$$\frac{C_p - C_v}{C_v} = 0.67$$

whereas for copper this ratio is 10^{-2} at 293 K. With solids, therefore, only a single quantity C is used: in practice, theoretical calculations are carried out at constant volume, while measurements are made at constant pressure.

Our aim is to relate the heat capacity to atomic structure with the help of models which are increasingly refined in order to achieve better and better agreement with experimental data. For simplicity, we first consider the special cause of a *monatomic* substance in which all the atoms play the same role (this is the situation in common metals).

The first model of atomic structure

Books dealing with the atomic structure of matter (e.g. SM,[†] pp. 57–9) lay much stress on the model of a crystal with its atoms at the nodes of a perfectly regular lattice. In reality, the nodes are merely the sites of the average atomic positions, around which the atoms themselves are vibrating due to their thermal agitation. The increase in internal energy of a crystal as the temperature rises originates from an increase in the amplitude of these vibrations, and it is precisely this that we wish to evaluate.

Let us first go back to the case of a perfect gas. Here, the atomic motion is completely disordered and atomic interaction is negligible, so that the thermal energy of the whole system is simply the statistical mean of the kinetic energy of the atoms. This mean energy is a measure of the temperature T (SM, p. 37), and the expression embodying the definition of T can then be written in the form:

$$<\tfrac{1}{2}mv^2> = 3k_BT/2$$

The coefficient k_B is **Boltzmann's constant**, with a value of 1.38×10^{-23} J K^{-1}.

An atom in a crystal vibrates about its equilibrium position. When the atom is displaced from this position, a restoring force is brought into play which does work corresponding to the reduction in the potential energy of the atom. The total energy of the atom is the sum of its mean kinetic and mean potential energies.

Take the simple case of a linear harmonic oscillator, with a point mass m subject to a displacement $x = A\sin \omega t$ and a restoring force $-Kx$. At any time t, the kinetic energy is

$$\tfrac{1}{2}mv^2 = \tfrac{1}{2}mA^2\omega^2\cos^2 \omega t$$

and the potential energy is

$$\tfrac{1}{2}Kx^2 = \tfrac{1}{2}KA^2\sin^2\omega t.$$

However, $\omega^2 = K/m$, so that the two mean values $\tfrac{1}{4}mA^2\omega^2$ and $\tfrac{1}{4}KA^2$ are equal: *the total energy of the oscillator is twice its mean kinetic energy*.

This result is still valid for an atom vibrating in three spatial dimensions about its equilibrium position. Moreover, whether the atom vibrates

[†]References given in this form are to Guinier, A. (1984). *The structure of matter*, Edward Arnold, London.

around a fixed point or suffers random displacements as in a gas, the mean value of its kinetic energy at a given temperature is the same: $3k_BT/2$.

In a crystal, therefore, the mean energy is:

for one atom: $u = 2(3/2)k_BT = 3k_BT$
for one mole: $U = N_A.3k_BT = 3RT$.

Avogadro's number N_A, is 6.02×10^{23} mol^{-1}, and the gas constant $R = N_Ak_B$ is 8.31 J K^{-1} mol^{-1}. It follows that the molar heat capacity is

$$C = dU/dT = 3R = 25 \text{ J K}^{-1} \text{ mol}^{-1}.$$

This is an important result, of a somewhat surprising simplicity.

So far, then, we have shown theoretically that:

for any monatomic crystal, the heat capacity is independent of temperature;
the molar heat capacity has a universal value which depends on Boltzmann's constant and which is the same for all chemical elements whatever the atomic mass or the crystal structure.

This is **Dulong and Petit's law**, which was established experimentally in 1819.

Justification of the theoretical calculation of heat capacity

We have used a very simple method for the derivation of these fundamental results, starting from the value for the kinetic energy of an atom in a perfect gas and extrapolating an elementary property of the linear harmonic oscillator. In fact, the calculation is rigorous since it relies on a general theorem of statistical mechanics known as the **principle of the equipartition of energy** or, more simply, as the **equipartition principle**. A precise statement of this principle and a proof of it is given in Box 1.

Box 1

The principle of the equipartition of energy

Take once again the example of the linear harmonic oscillator with mass m, restoring force constant K, frequency[†] $\omega = \sqrt{K/m}$, displacement $x = A\sin\omega t$. The constant total mechanical energy is the sum of

[†]Strictly speaking, the frequency is $\omega/2\pi$ and ω is the angular frequency or pulsatance. In this book, we adopt the abbreviated expression which is now normal among those specializing in the field.

the potential and kinetic energies and is therefore:

$$E(x, v) = E_p(x) + E_k(v) = \tfrac{1}{2}Kx^2 + \tfrac{1}{2}mv^2 = \tfrac{1}{2}KA^2.$$

Take a whole assembly of identical oscillators. The variables x and v, defining the position and velocity of the particle at a given time, fix the state of an oscillator determined either by its amplitude A or by its energy $\tfrac{1}{2}KA^2$.

Boltzmann's statistical thermodynamics predicts the distribution of the states of the oscillators in the system and thus the probability that an oscillator is described by the pair of variables x and v to within Δx and Δv. This probability is proportional to:

$$\exp[-E(x, v)/k_B T)]\Delta x \Delta v.$$

This statistical law is a *quantitative* expression of the idea of thermal agitation. Thus, when the temperature tends to infinity, the exponent tends to zero and all the energy states become equally probable (this is complete disorder). On the other hand, when the temperature tends to zero, all the probabilities become negligible except that of the state of absolute rest, $E = 0$, which corresponds to $x = 0, v = 0$.

Given this probability law, any macroscopic physical quantity characteristic of the system is calculated as an average, weighted according to the probabilities, i.e. as the sum of the possible values each multiplied by the probability of their occurrence. Thus, the average energy of an oscillator is calculated from:

$$\bar{E} = \frac{1}{Z} \iint E(x, v) \exp\left[-\frac{E(x, v)}{k_B T}\right] dx\, dv.$$

Z is a factor required to 'normalize' the probabilities, since the sum of the probabilities over all states must be unity, i.e.

$$Z = \iint \exp\left[-\frac{E(x, v)}{k_B T}\right] dx\, dv.$$

Since the energy E is the sum of two terms, E_p and E_k, we have that:

$$E = \bar{E}_p + \bar{E}_k,$$

where:

$$\bar{E}_p = \frac{1}{Z} \iint \frac{1}{2} K x^2 \exp\left[-\frac{E(x, v)}{k_B T}\right] dx\, dv,$$

$$\bar{E}_k = \frac{1}{Z}\iint \frac{1}{2} m v^2 \exp\left[-\frac{E(x, v)}{k_B T}\right] dx\, dv.$$

Taking the case of $\bar{E}_p$, the double integral amounts to the product of two single integrals:

$$\bar{E}_p = \frac{1}{Z}\int_{-\infty}^{+\infty} \frac{1}{2} K x^2 \exp\left(-\frac{1}{2}\frac{K x^2}{k_B T}\right) dx \int_{-\infty}^{+\infty} \exp\left(-\frac{1}{2}\frac{m v^2}{k_B T}\right) dv$$

and the same for Z:

$$Z = \int_{-\infty}^{+\infty} \exp\left(-\frac{1}{2}\frac{K x^2}{k_B T}\right) dx \int_{-\infty}^{+\infty} \exp\left(-\frac{1}{2}\frac{m v^2}{k_B T}\right) dv.$$

Substituting the expression for Z into that for $\bar{E}_p$, it is clear that the integrals with respect to velocity cancel, leaving only those with respect to position:

$$\bar{E}_p = \frac{\displaystyle\int_{-\infty}^{+\infty} \frac{1}{2} K x^2 \exp\left(-\frac{1}{2}\frac{K x^2}{k_B T}\right) dx}{\displaystyle\int_{-\infty}^{+\infty} \exp\left(-\frac{1}{2}\frac{K x^2}{k_B T}\right) dx}.$$

The same calculation can be carried out for $\bar{E}_k$, and a similar expression is obtained with $\frac{1}{2}Kx^2$ replaced by $\frac{1}{2}mv^2$ and dx by dv.

The values of these integrals can be found from standard tables and yield the expressions:

$$\bar{E}_p = \bar{E}_k = \frac{\dfrac{K}{2}\dfrac{\sqrt{\pi}}{2}\left(\dfrac{2k_B T}{K}\right)^{\frac{3}{2}}}{\sqrt{\pi}\left(\dfrac{2k_B T}{K}\right)^{\frac{1}{2}}} = \frac{k_B T}{2}.$$

The equality of $\bar{E}_p$ and $\bar{E}_k$ arises from the fact that they are both quadratic functions of x or v.

We therefore have that:

$$\bar{E} = \bar{E}_p + \bar{E}_k = 2(\tfrac{1}{2}k_B T) = k_B T.$$

This result can be generalized to the case where there is a large number of variables $q_1, q_2, \ldots$ and *where the energy is the sum of terms of the type $a_i q_i^2$ only:*

$$E(q_1, q_2, \ldots) = \sum_i a_i q_i^2.$$

As an example, for an atom vibrating around a fixed site:

$$E = \frac{1}{2} K(x^2 + y^2 + z^2) + \frac{1}{2} m(v_x^2 + v_y^2 + v_z^2).$$

Each term contributes $\frac{1}{2}k_B T$ to the total average energy and this is what is called the **equipartition of energy**. We have:

$$\bar{E} = n_d(\tfrac{1}{2}k_B T)$$

where n_d is the number of **degrees of freedom**, i.e. the *number of independent quadratic variable terms entering into the expression for the energy*. It should be pointed out that this number may be less than the number of independent variables defining the state of the system. Thus, in a perfect gas, the state of a particle is defined by its position and its velocity, but only the velocity variables occur in the expression for the energy.

For a perfect gas in three dimensions:

$$n_d = 3, \bar{E} = 3k_B T/2.$$

For a three-dimensional harmonic oscillator:

$$n_d = 6, \bar{E} = 3k_B T.$$

Comparison of theory with experiment

1 Points of agreement

For many chemical elements (Table 1.1 gives a few examples), the heat capacity shows very little variation over a large temperature range and, except for a few such as Be or C, the experimental values of the molar heat capacity generally differ from the theoretical value by less than 5 per cent. This is clearly a distinct success for the extremely simple theory we have used.

When Dulong and Petit's law is obeyed, the heat capacity of a given mass of metal depends only on the number of atoms it contains. It does not depend on the atomic structure: thus, liquid mercury has the same heat capacity as solid mercury at the melting point.

The historical importance of this law arises from the fact that it is directly related to the existence of atoms. By measuring the heat capacity of unit mass of an element (i.e., the *specific heat capacity*, 25/M in J kg^{-1} K^{-1}), an

Table 1.1 Molar heat capacity of several elements. (Theoretical value according to Dulong and Petit's law: $25\,\mathrm{J\,K^{-1}\,mol^{-1}}$).

	T		
Element	300 K	500 K	1000 K
Be	16.4	22.2	
C (diamond)	6.24	13.4	21.6
Al	24.3	25.8	
Cu	24.2	26.1	28.9
Pb	26.8	29.4	

approximate value for its molar mass M is obtained. Consider gold and copper, for example: the inverse ratio of their specific heat capacities at high temperature is 0.38/0.13 or 2.9. This gives an approximate value for the ratio of their atomic masses whose exact value is 197/63 = 3.1.

2 *Points of disagreement*

Although the heat capacity of elements such as Cu and Pb varies very little above room temperature, it decreases at low temperatures and for all substances *it becomes zero at the absolute zero*. For copper, for example, there is a pronounced decrease below 150 K (Fig. 1.1). Furthermore, there are substances for which the molar heat capacity even at room temperature is already well below the theoretical value, e.g. diamond has a value of only 6 instead of $25\,\mathrm{J\ K^{-1}\ mol^{-1}}$.

Nothing in the model we have used enables us to explain this behaviour, which is completely different from what is predicted. Such a large disagreement can only be resolved by introducing a new idea into the over-simplified theory. What we need is a modification in the low temperature region which leaves the results intact at medium and high temperatures, where the agreement with experiment is remarkably good.

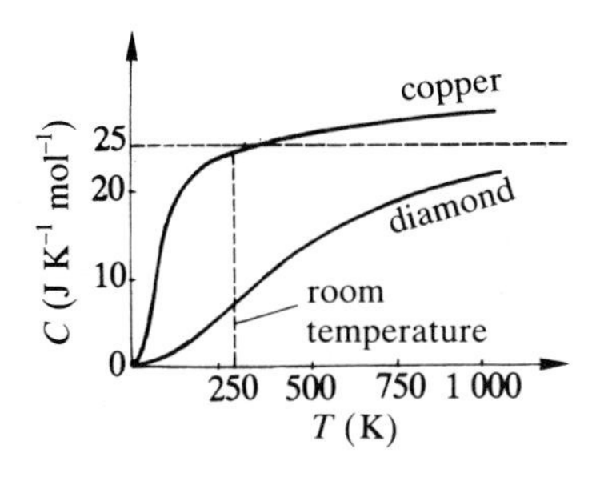

Fig. 1.1. Variation with temperature of the molar heat capacity of copper and diamond.

The second model: Einstein's model

We first look in more detail at what we have until now called atomic vibrations. In a crystal, particularly a metal, the atoms are attracted to each other by cohesive forces. In addition, since they cannot penetrate each other, a repulsive force comes into play when their centres approach too closely. The true interatomic distance is the separation at which these two forces exactly balance each other. Suppose that all the atoms in a crystal are at their equilibrium positions. If an atom is displaced, it experiences a restoring force which, to a first approximation, is proportional to the distance $\Delta\boldsymbol{x}$ from the equilibrium position (this is known as the harmonic approximation), i.e. $\boldsymbol{F} = -K\Delta\boldsymbol{x}$. This is analogous to the force on a mass suspended from a spring.

Remember that the vibration frequency is $\omega = \sqrt{(K/m)}$. To find an order of magnitude for the vibration frequency of the atoms in a crystal, the value of the force constant K must be known, and this is derived from the macroscopic elastic properties of the metal.

For iron, Young's modulus E (see p. 178) is 20×10^{10} N m^{-2}. This means that the force which extends the length l of a rod of unit cross-section elastically by Δx is $E\Delta x/l$. Now an atomic plane in iron contains $n = 2.5 \times 10^{19}$ atoms per m^2 and the distance between two adjacent planes is $l = 3 \times 10^{-10}$m. The macroscopic force applied to the rod is equivalent to a force F per atom which would be related to Δx by the expression:

$$F = \frac{1}{n}\frac{E}{l}\,\Delta x = K\Delta x.$$

The force constant K is thus given by E/ln and, since the mass of an iron atom is $m = 9 \times 10^{-26}$kg, the frequency of atomic vibrations is:

$$\omega = \sqrt{(K/m)} = 1.7 \times 10^{13}\,\text{Hz}.$$

This is a very approximate calculation and all that we shall assume from it is that the frequency of atomic vibrations in solids has an order of magnitude of between 10^{12} and 10^{14} Hz.

The atom vibrates in space: its motion can be regarded as the resultant of three linear oscillations along three perpendicular axes. This is the basis of Einstein's model: *the motion of the N_A atoms in a mole of a crystal is represented by the motion of $3N_A$ identical oscillators having the same frequency and not interacting with each other.* These oscillators form the reservoir of energy which stores the heat supplied to the crystal: when it is heated, the oscillator amplitudes increase; when it is cooled, the amplitudes decrease.

The new idea contributed by Einstein was that of **quantization.** In classical physics, the total energy of an oscillator of mass m, amplitude A and frequency ω is $\frac{1}{2}m\omega^2A^2$. It can therefore vary continuously since A may take any value.

Quantum theory, on the other hand, only allows the energy of the oscillator to have one of a set of discontinuous values and any change in energy can only be by an integral number of energy **quanta**. The elementary quantum of energy is $\hbar\omega$, where $\hbar$ is Planck's constant divided by 2π:

$$\hbar = 1.054 \times 10^{-34}\ \mathrm{J\ s}.$$

The oscillator therefore has a set of **energy levels** which can be allocated numbers according to the number of quanta they contain.

Later on, quantum mechanics retained the quantization of the energy changes in the oscillator but made one addition to Einstein's model: the minimum energy of the oscillator at the absolute zero cannot itself be zero. If it were, Heisenberg's uncertainty principle would be violated, since the vibrating mass would be stationary and its exact position and velocity would be known simultaneously. This means that there is a **zero-point energy**, and quantum mechanical calculations give this the value $\frac{1}{2}\hbar\omega$. The nth energy state of the oscillator thus has an energy:

$$E_{\mathrm{n}} = (n + \tfrac{1}{2})\hbar\omega.$$

However, this extra term does not affect the theory of the heat capacity.

When the energy, and thus the amplitude, of the oscillator is large enough, the fact that the possible levels are discrete is not significant since the energy quantum is then much smaller than the experimental error in making the most refined measurements. This is so for any macroscopic oscillators that we could possibly construct. It is also true for atomic oscillators when the temperature of the solid is sufficiently high. Under those conditions, therefore, the Einstein model is almost identical with our first classical model, whose predictions are verified by experiment and continue to be valid.

At temperatures close to the absolute zero, on the other hand, the oscillator energies are very small and the replacement of continuous variations by abrupt changes from one level to another can have appreciable effects. It is because of this that the quantum model of Einstein is able to explain the behaviour of the heat capacity at low temperatures, a phenomenon that was incomprehensible in terms of classical physics. It was the very success achieved by Einstein in this problem that formed one of the first proofs of the existence of energy quanta.

Consider now the set of $3N$ oscillators, corresponding to N atoms, in equilibrium at temperature T. According to classical theory, their average energy is $k_B T$. According to quantum theory, each contains a number of quanta given by $(n + \frac{1}{2})$, which may be different for different oscillators. The mean number of quanta about which the value of n fluctuates is given as a function of temperature by Bose–Einstein statistics.

Box 2

The Bose–Einstein distribution

Thermodynamic arguments lead to the following result: the probability that an oscillator at temperature T will be occupying the energy state $E_n = (n + \frac{1}{2})\hbar\omega$ labelled by its quantum number n is proportional to $\exp(-E_n/k_B T)$. The mean value of n at temperature T can then be calculated using the method of Box 1, except that here the expressions are discrete sums and not integrals:

$$\bar{n} = \frac{1}{Z} \sum n \exp\left(-\frac{E_n}{k_B T}\right) = \frac{\exp\left(-\frac{1}{2}\frac{\hbar\omega}{k_B T}\right)}{Z} \sum n \exp\left(-\frac{n\hbar\omega}{k_B T}\right)$$

$$Z = \sum \exp\left(-\frac{E_n}{k_B T}\right) = \exp\left(-\frac{1}{2}\frac{\hbar\omega}{k_B T}\right) \sum \exp\left(-\frac{n\hbar\omega}{k_B T}\right)$$

Denote $\exp(-\hbar\omega/k_B T)$ by y : $\bar{n}$ can then be expressed as the ratio of two sums:

$$\bar{n} = \Sigma n y^n / \Sigma y^n.$$

The denominator is the sum of a simple geometric series with ratio y:

$$\Sigma y^n = 1 + y + y^2 + \ldots = 1/(1 - y).$$

The numerator is the following sum:

$$\Sigma n y^n = y + 2y^2 + 3y^3 + \ldots = y(1 + 2y + 3y^2 + \ldots)$$

and this is simply the derivative of the previous sum multiplied by y, so that:

$$\Sigma n y^n = y/(1 - y)^2$$

and hence: $$\bar{n} = y/(1 - y) = \frac{1}{\exp(\hbar\omega/k_B T) - 1}.$$

The mean value $\bar{n}$ gives the average energy of the set of 3N oscillators:

$$\bar{E} = 3N(\bar{n} + \tfrac{1}{2})$$

or

$$\bar{E} = 3N\left(\frac{\hbar\omega}{2} + \frac{\hbar\,\omega}{\exp\left(\dfrac{\hbar\,\omega}{k_B T}\right) - 1}\right)$$

What is the significance of this expression? It depends only on the one parameter $\hbar\omega/k_B T$. If a *characteristic temperature* Θ_E, such that $\hbar\omega = k_B\Theta_E$, is introduced, then we obtain a universal formula valid for any solid (i.e. for any oscillator frequency) in terms of the 'reduced' temperature T/Θ_E:

$$\bar{E} = 3\,R\,\Theta_E\left(\frac{1}{2} + \frac{1}{\exp\left(\dfrac{\Theta_E}{T}\right) - 1}\right).$$

The molar heat capacity is obtained by differentiating the mean energy $\bar{E}$ with respect to T. We then obtain:

$$C = \frac{d\bar{E}}{dT} = 3\,R\left(\frac{\Theta_E}{T}\right)^2 \frac{\exp\dfrac{\Theta_E}{T}}{\left(\exp\left(\dfrac{\Theta_E}{T}\right) - 1\right)^2}.$$

This function has a 'sigmoid' shape (Fig. 1.2). It starts from zero when $T = 0$; it increases as T increases, slowly at first; it reaches the value 1.49R

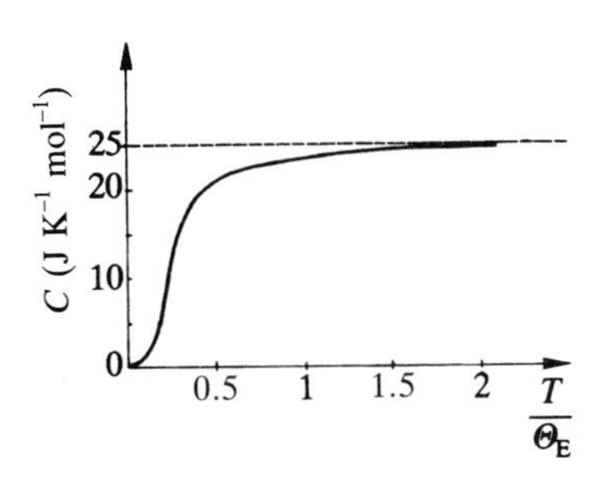

Fig. 1.2. The molar heat capacity according to Einstein's theory.

for $T = \Theta_E/3$ and then tends towards the classical value $3R$. Above $T = 1.3\Theta_E$, the function has the value $3R$ to within 5%.

At very low temperatures, a significant proportion of the oscillators are still in the ground state. A rise in temperature of ΔT only excites a small number, so that the amount of heat absorbed is well below the classical value $3R\Delta T$: the heat capacity remains small.

The theoretical Einstein curve is very similar to many experimental curves. For example, the best agreement in the case of copper (Fig. 1.3(a)) is obtained by taking Θ_E equal to 244 K, which corresponds to an oscillator frequency of 3.2×10^{13} Hz.

However, a clear divergence can be seen near zero: the experimental curve runs very much above the theoretical curve (Fig. 1.3(b)).

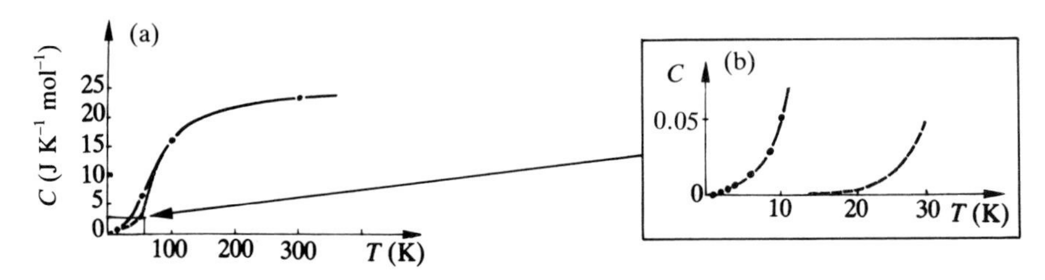

Fig. 1.3. (a) The molar heat capacity of copper as a function of temperature: (•) experimental points, (----) theoretical Einstein curve with $\Theta_E = 244$ K, (——) theoretical Debye curve with $\Theta_D = 315$ K. (b) The same curves over the range 0–30 K.

For diamond, the agreement is satisfactory near room temperature if Θ_E is taken to be 2000 K. Diamond does not obey Dulong and Petit's law at ordinary temperatures precisely because the characteristic temperature is so much higher. Einstein's theory explains why this occurs: diamond is very hard with strong cohesive forces. The restoring force constant is therefore large and the frequency of atomic vibrations very high, and it follows that the characteristic temperature is also very high. Copper, on the other hand, is quite a soft metal: its vibration frequency is lower and its characteristic temperature is below room temperature.

The value and the importance of Einstein's theory are demonstrated by the close relationship it reveals between the elastic constant and the heat capacity of a crystal, two apparently unconnected quantities. Such relationships cannot be chance coincidences: they show there is a valid 'kernel' in the theory which is an expression of reality. This kernel will persist during later refinements of the theory.

There are, nevertheless, serious objections to Einstein's theory, not simply because of disagreement with experiment (which is on the whole quite minor) but mostly because it rests on an assumption that is undoubtedly incorrect: the independence of the motion of neighbouring atoms. It was to take into account the interactions between the motion of the atoms in a solid that Debye proposed the third model of thermal agitation in crystals.

Elastic waves in a crystal: the Debye model

We first review the classical results on **longitudinal** vibrations in a rod. The creation of a small local deformation (or strain) directed along the axis of the rod produces stresses in the same direction because of the elasticity of the material. These stresses produce strains in adjacent sections which, in their turn, react on their neighbours and so on. The cohesion and elasticity of the rod thus cause the deformation to be *propagated* along its length with a velocity v determined by the properties of the material of which the rod is made (Young's modulus and density).

We now suppose that, in place of the short pulse, we impose a steady longitudinal sinusoidal vibration of frequency ω on a section of the rod. Interactions in the solid cause this local vibration to be propagated in such a way that all the sections vibrate with the same frequency and amplitude. However, the vibrations at two points separated by a distance l are out of phase because of the time l/v taken for propagation (angular phase difference $\omega l/v$). Points separated by a distance $\lambda = v(2\pi/\omega)$ or vT, T being the period of vibration, vibrate in phase. At a given instant of time, the displacements along the rod are represented by a sine curve with wavelength λ. As time passes, the sine curve moves along the rod at the velocity v giving rise to a **progressive wave** characterized by its amplitude,

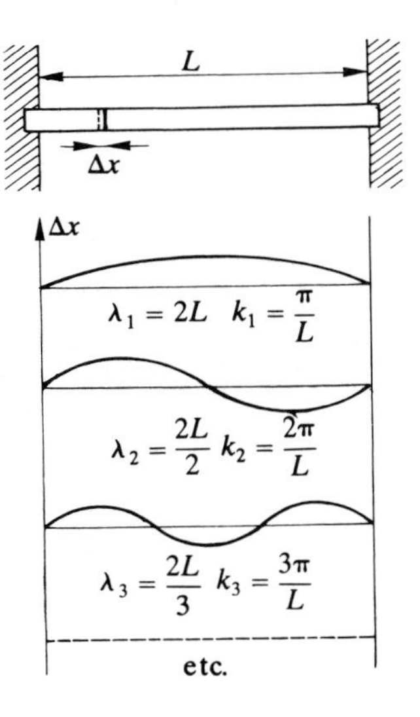

Fig. 1.4. Normal modes of the longitudinal vibrations of a rod fixed at both ends.

its frequency and either its wavelength or, more commonly, its **wave vector** $k = 2\pi/\lambda = \omega/v$.

If the rod is fixed at one end, the enforced immobility at this point causes the progressive wave to be reflected. If a rod of length L is fixed at both ends, these boundary conditions determine the **normal modes** of vibration of the rod, i.e. stationary waves whose wavelengths form a series of discrete values $\lambda_n = 2L/n$, *where n is an integer* (Fig. 1.4, which is reminiscent of the bow-like shapes of the vibrating string in Melde's experiment). Another way of putting this is to say that the wave vectors of the normal modes are integral multiples of a minimum vector π/L, so that:

$$k_n = n\pi/L.$$

The frequency corresponding to the nth mode is $\omega_n = vk_n$.

Normal modes are essential for describing the vibrations of the rod: any actual motion can be analysed into a superposition of normal modes with suitable amplitudes and phases.

We now transfer these one-dimensional results to the case of a solid, e.g. a cube of side L. An elastic wave is defined by its frequency, its wave vector and the vector amplitude of the vibration. The direction of the wave vector is normal to the plane of the wavefront, while its magnitude is $2\pi/\lambda$. The cube also has its normal modes of vibration: in other words, the wave vectors can only possess certain discrete values for their magnitudes and directions. We state the result that the projections of the allowed wave vectors on to axes parallel to the sides of the cube are $n_1(\pi/L)$, $n_2(\pi/L)$, $n_3(\pi/L)$, where n_1, n_2, and n_3 are *integers*. This is a generalization of the result quoted above for vibrating rods.

If we consider only those wavelengths that are large compared with atomic dimensions, the discontinuous nature of the material can be neglected and it can be described by its macroscopic properties (its density and elastic constants). Thus, for longitudinal waves in the acoustic and ultrasonic ranges ($\lambda > 1\,\mu m$), the velocity of propagation is independent of the wavelength and is equal to the speed of sound in the solid (of the order of $5000\,m\,s^{-1}$).

However, when we wish to deal with the general case of vibrations in a crystal, the normal modes have to be calculated from the interatomic forces by taking into account the atomic structure of the crystal. It can well be imagined that such a calculation is extremely complex. Not only that, but the present state of our knowledge does not enable us to carry out the calculation using data on the types of atoms in the crystal and their positions alone: parameters derived from macroscopic measurements also have to be introduced.

We shall be content with an account of those results from crystal dynamics which are of the greatest importance as regards their consequences and generality. Before that, however, we deal with the case of a linear chain of atoms (a one-dimensional lattice) in Box 3 below, an example in which only simple methods of calculation are needed. By comparing the results from this example with those of the classic case of the vibrating homogenous rod, the effects caused by the discontinuous nature of the solid are clearly revealed.

Box 3

Longitudinal vibrations in a chain of identical atoms connected by springs

When the chain (Fig. 1.5) is in motion, each atom is described by its displacement $u_n(t)$ at time t with respect to its equilibrium rest position $x_n = na$, where n is an integer and a is the lattice spacing. The nth mass is subject to two forces:

- one from the spring on the left, whose extension of algebraic value $u_n - u_{n-1}$ is caused by the nth and $(n-1)$th masses;
- one from the spring on the right, whose extension of algebraic value $u_n - u_{n+1}$ is caused by the nth and $(n+1)$th masses.

The resultant force is thus:

$$f = -K(u_n - u_{n-1} + u_n - u_{n+1})$$
$$= -K(2u_n - u_{n-1} - u_{n+1}).$$

The equation of motion of the nth mass is thus:

$$m\,\mathrm{d}^2u_n/\mathrm{d}t^2 = -K(2u_n - u_{n-1} - u_{n+1}).$$

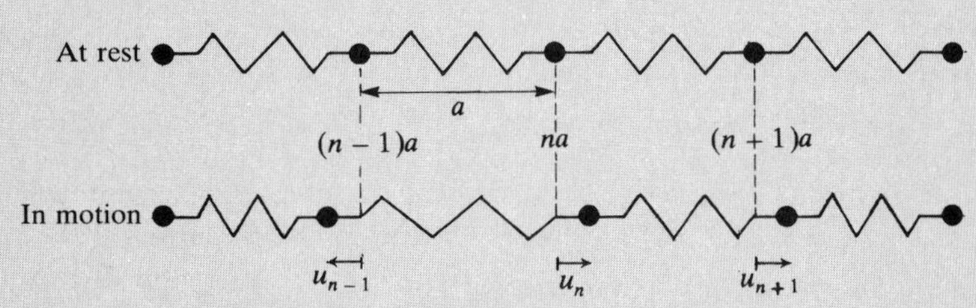

Fig. 1.5. Vibrations of a linear chain of atoms.

The difficulty arises from the presence of the positions of the $(n-1)$th and $(n+1)$th atoms in the equation of motion of the nth atom. It is these **coupling** terms which cause the motion of one mass to be transmitted to neighbouring masses and thus to the whole chain.

We next attempt to find a solution in the form of a progressive wave:

$$u_n = u_0 \cos(\omega t - kna).$$

The parameter u_n is interpreted as the longitudinal displacement which would be produced by a wave of amplitude u_0, frequency ω and wave vector k at the points $x_n = na$ where the masses in the chain at rest are located. We substitute this expression in the equation of motion to see under what conditions that is possible. If we use the identity

$$\cos[\omega t - k(n-1)a] + \cos[\omega t - k(n+1)a] = 2 \cos ka \cos(\omega t - kna)$$

we obtain the equation:

$$-m\omega^2 u_0 = -K(2 - 2\cos ka)u_0 = -4K \sin^2(ka/2)\, u_0.$$

This equation only has a non-zero solution for u_0 if:

$$m\omega^2 = 4K\sin^2(ka/2)$$

or (Fig. 1.6):

$$\omega = 2\sqrt{\frac{K}{m}}\left|\sin\frac{ka}{2}\right|$$

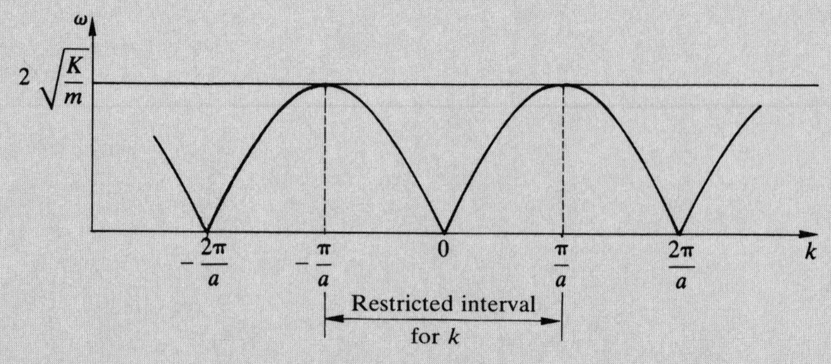

Fig. 1.6. Dispersion curve for the linear chain: ω, the vibration frequency, plotted against the wave vector $k = 2\pi/\lambda$.

Consequences of the expression $\omega = f(k)$

For long wavelengths, or small k, the above expression can be written:

$$\omega \approx a\sqrt{\frac{K}{m}}\,k$$

i.e. the velocity of propagation is $\upsilon = a\sqrt{K/m}$ and is thus independent of the wavelength. Now for a homogeneous rod, the velocity of propagation is $\upsilon = \sqrt{E/\mu}$, where E is Young's modulus and μ is the density. *The two expressions are equivalent.* In fact, if S is the cross-section of the rod and Δa the increment in the interatomic distance a due to the effect of a force F, we can write for the chain and the rod:

$$F = K\Delta a \quad \text{and} \quad F/S = E(\Delta a/a).$$

Furthermore, so as to give the atomic chain and the rod the same mass per unit length, μS must be put equal to m/a. From these relationships, we deduce that:

$$E/\mu = Ka^2/m.$$

At short wavelengths, say less than $10a$, the frequency ω is no longer proportional to k: the velocity of propagation decreases as k increases. **Dispersion** occurs, and this is a consequence of the atomic structure of the chain.

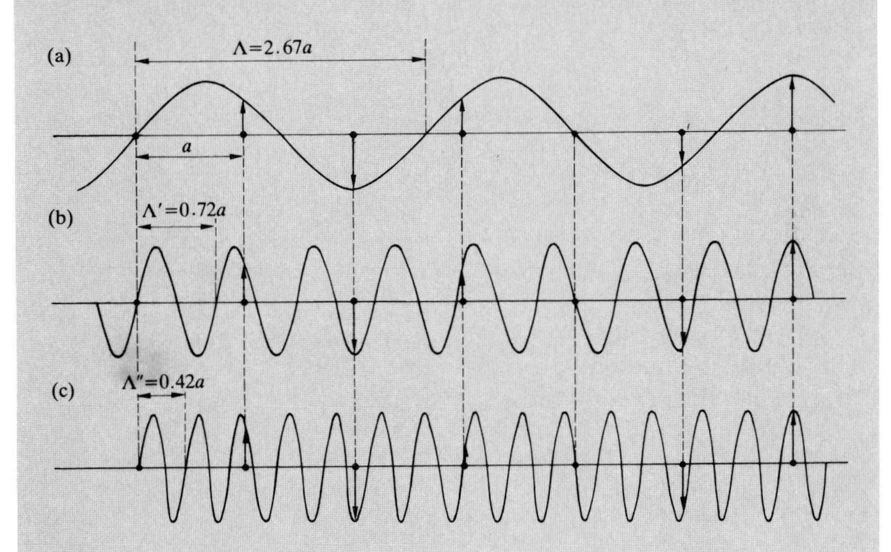

Fig. 1.7. The displacements of all the atoms in a chain of period a are the same for all wave vectors of the form $k+2\pi p/a$. (a) $p = 0$; $\Lambda = 2.67a$; $k/2\pi = 0.375/a$; (b) $p = 1$; $\Lambda' = 0.72a$; $k'/2\pi = k/2\pi + 1/a = 1.375/a$; (c) $p = 2$; $\Lambda'' = 0.42a$; $k''/2\pi = k/2\pi + 2/a = 2.375/a$.

The vibration frequency becomes a maximum when $k = \pi/a$. The wave is then represented by the equation:

$$u_n = u_0\cos(\omega t - n\pi) = u_0(-1)^n\cos\omega t.$$

Successive atoms have amplitudes of alternating sign. This wave can be considered as a stationary wave (constant phase, and amplitude varying from one atom to the next). Such a wave cannot transport energy (the group velocity $d\omega/dk$ is zero).

There is another, and very fundamental, consequence of the discontinuity on the atomic scale: there is a lower limit to the wavelength, i.e. an upper limit to the wave vector. This arises from the fact that the atoms form a regular lattice and that the elastic wave has a physical meaning only at points along it with coordinates equal to na. Suppose there is a wave vector k and another $k' = k + 2\pi p/a$, where p is an integer. The displacement of the nth atom for the wave k' is:

$$u_n = u_0\cos\left[\omega t - \left(k + \frac{2\pi p}{a}\right)n a\right] = u_0\cos(\omega t - kna - 2\pi np)$$

i.e.

$$u_n = u_0\cos(\omega t - kna).$$

The displacement u_n is the same as for the wave vector k, for any atom whatsoever (Fig. 1.7). Thus the wave k' cannot be physically differentiated from the wave k. Hence, all the normal modes of the chain can be described simply by considering only those waves whose wave vector lies in an interval of magnitude $2\pi/a$, e.g. those for which k lies in the interval $[-\pi/a, \pi/a]$ (Fig. 1.6) and are thus being propagated in both directions.

In addition, the boundary conditions (arising from the fact that the chain must be finite in length and contain only N atoms) force the number of independent solutions, and thus the number of k values in the interval $[-\pi/a, \pi/a]$, to be finite and equal to N. The exact positions of these k values depend on the imposed conditions, but *their number is always N.*

Characteristics of the normal modes of vibration of a crystal

The main results from similar calculations of the modes of vibration of a crystal are listed below without mathematical justification. Their meaning and significance can be appreciated by comparing them with the complete treatment of the linear chain given in Box 3.

1 Number of modes

The modes of vibration of a crystal form a discrete and finite series. For a crystal containing N atoms, whatever the external shape, there are N *wave vectors* giving distinct modes. Three directions of the vector amplitude correspond to each of these modes: *the total number of modes is therefore 3N.*

In the linear chain (Box 3), we counted N longitudinal vibrational modes. However, the chain can also propagate transverse vibrations which can be resolved into two vibrations in perpendicular directions. This clearly means that there are three waves per wave vector. For crystals, the three amplitude vectors form an orthogonal triad: if the wave vector is directed along a symmetry axis of the crystal, one wave is longitudinal and the two others are transverse. For a general direction, this arrangement is only approximate.

2 Upper limit for the wave vector

For the linear chain of period a, the wave vectors corresponding to physically distinct waves are confined to the interval $[-\pi/a, \pi/a]$. There is a similar limitation on the wave vectors for a crystal: the ends of the wave vectors lie inside a polyhedral surface known as a **Brillouin zone**, defined geometrically in terms of the crystal lattice. A description of these zones is given in Box 4.

Box 4

Definition of the reciprocal lattice and construction of Brillouin zones

For elastic waves propagated in a crystal, we shall generalize the limits for the wave vectors established for the linear chain ($-\pi/a < k \leqslant \pi/a$).

In the linear chain, the atomic displacement is given by:

$$u_n = u_0 \cos(\omega t - kx_n)$$

with $x_n = na$.

For the wave in the crystal (limiting ourselves to the case of a simple crystal containing one atom per unit cell), the expression becomes:

$$\mathbf{u}_{n_a n_b n_c} = \mathbf{u}_0 \cos(\omega t - \mathbf{k} \cdot \mathbf{r}_{n_a n_b n_c}).$$

The vector $\boldsymbol{u}_{n_a n_b n_c}$ is the displacement of the atom with coordinates n_a, n_b, n_c at the site defined by the vector:

$$\boldsymbol{r}_{n_a n_b n_c} = n_a \boldsymbol{a} + n_b \boldsymbol{b} + n_c \boldsymbol{c}$$

a*, *b*,** and ***c being the basis vectors of the unit cell, and the coordinates n_a, n_b, and n_c being integers.

Just as the product kx_n determined the phase at the atomic site in the linear case, so here does the scalar product $\mathbf{k}\cdot\mathbf{r}_{n_a n_b n_c}$ at the site (n_a, n_b, n_c). A simple expression for the scalar product can be obtained if the vectors $\boldsymbol{k}$ are referred to three basis vectors $\boldsymbol{a}^*$, $\boldsymbol{b}^*$, $\boldsymbol{c}^*$ defining the **reciprocal lattice** of the crystal. These vectors are defined by the following relationships:[†]

$$\begin{array}{lll} \boldsymbol{a}^*\cdot\boldsymbol{a} = 2\pi & \boldsymbol{b}^*\cdot\boldsymbol{a} = 0 & \boldsymbol{c}^*\cdot\boldsymbol{a} = 0 \\ \boldsymbol{a}^*\cdot\boldsymbol{b} = 0 & \boldsymbol{b}^*\cdot\boldsymbol{b} = 2\pi & \boldsymbol{c}^*\cdot\boldsymbol{b} = 0 \\ \boldsymbol{a}^*\cdot\boldsymbol{c} = 0 & \boldsymbol{b}^*\cdot\boldsymbol{c} = 0 & \boldsymbol{c}^*\cdot\boldsymbol{c} = 2\pi. \end{array}$$

These relationships express the fact that the vector $\boldsymbol{a}^*$ is normal to the crystal planes defined by the vectors $\boldsymbol{b}$ and $\boldsymbol{c}$, and also that the magnitude of $\boldsymbol{a}^*$, apart from the factor 2π, is the reciprocal of the interplanar spacing of the ($\boldsymbol{b}$, $\boldsymbol{c}$) planes.

This property can be generalized: using the basis vectors $\boldsymbol{a}^*$, $\boldsymbol{b}^*$, and $\boldsymbol{c}^*$, the reciprocal lattice $\boldsymbol{r}^*_{hkl} = h\boldsymbol{a}^* + k\boldsymbol{b}^* + l\boldsymbol{c}^*$ can be

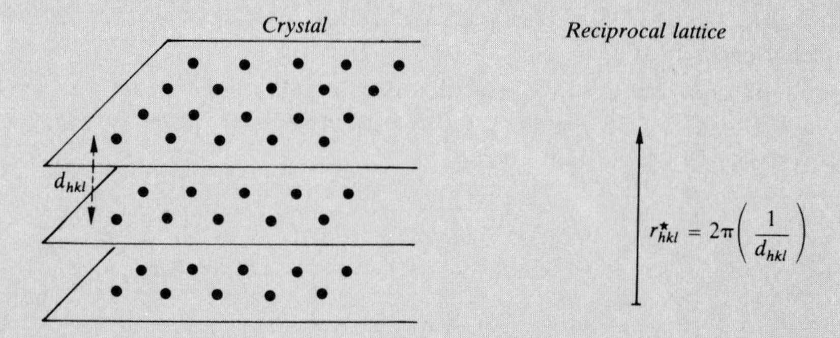

Fig. 1.8. Correspondence between the lattice planes of a crystal and reciprocal lattice vectors.

[†]The solution of this system of equations takes the form $\boldsymbol{a}^* = (2\pi/V)\boldsymbol{b} \times \boldsymbol{c}$, etc, where V is the unit cell volume given by $\boldsymbol{a}\cdot(\boldsymbol{b} \times \boldsymbol{c})$.

constructed, where h, k, and l are integers. Any vector in this reciprocal space is normal to a family of lattice planes of the crystal and its length, apart from the factor 2π, is the reciprocal of the interplanar spacing of this family of planes (Fig. 1.8).

A vector $\boldsymbol{k}$ being defined by its projections on the axes of the reciprocal lattice ($\boldsymbol{k} = p\boldsymbol{a}^* + q\boldsymbol{b}^* + r\boldsymbol{c}^*$), and a vector $\mathbf{r}_{n_a n_b n_c}$ being defined in the crystal lattice ($\boldsymbol{r}_{n_a n_b n_c} = n_a\mathbf{a} + n_b\boldsymbol{b} + n_c\mathbf{c}$), the defining relations for $\boldsymbol{a}^*$, $\boldsymbol{b}^*$, and $\boldsymbol{c}^*$ can be used to prove that the scalar product of $\boldsymbol{k}$ and $\boldsymbol{r}$ is

$$\boldsymbol{k}\cdot\boldsymbol{r}_{n_a n_b n_c} = 2\pi(pn_a + qn_b + rn_c).$$

This has the same simple form as in an orthonormal system but the expression is valid for any crystal lattice. Let us associate with the elastic wave of wave vector $\boldsymbol{k}$ all the waves with wave vector $\boldsymbol{k}' = \boldsymbol{k} + \boldsymbol{r}^*_{hkl}$, where $\boldsymbol{r}^*$ is a vector of the reciprocal lattice. It is found that:

$$\boldsymbol{k}'\cdot\boldsymbol{r}_n = \boldsymbol{k}\cdot\boldsymbol{r}_{n_a n_b n_c} + \boldsymbol{r}^*_{hkl}\cdot\boldsymbol{r}_{n_a n_b n_c},$$

i.e. that

$$\boldsymbol{k}'\cdot\boldsymbol{r}_n = \boldsymbol{k}\cdot\boldsymbol{r}_{n_a n_b n_c} + 2\pi(hn_a + kn_b + ln_c).$$

Thus, $\boldsymbol{k}'\cdot\boldsymbol{r}_{n_a n_b n_c}$ is equal to $\boldsymbol{k}\cdot\boldsymbol{r}_{n_a n_b n_c}$ apart from an *integral* multiple of 2π. The phases of the two waves are identical over all atoms. *The two waves are physically indistinguishable* since the displacement u_n of each of the atoms is the same. It follows that the study of elastic waves can be limited to those whose wave vectors lie within a certain volume of the reciprocal lattice, the **Brillouin zone**. This is a region such that any point within it is nearer the origin than any other node of the reciprocal lattice.

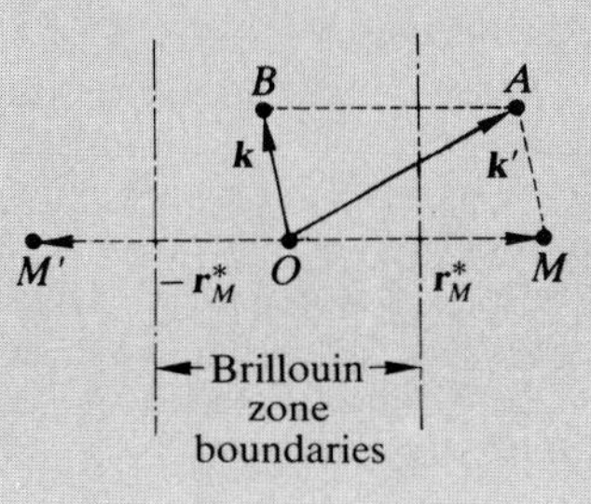

Fig. 1.9. All modes of vibrations can be represented by wave vectors located inside the Brillouin zone.

To demonstrate this, suppose that the end of A of a vector $\boldsymbol{k} = \overrightarrow{OA}$ is nearer a node M than the origin. We can replace $\boldsymbol{k}$ by the vector $\boldsymbol{k}'$ $= \overrightarrow{OB} = \boldsymbol{k} - \boldsymbol{r}^*{}_{\mathrm{M}}$. The point B is nearer the origin than the node M (Fig. 1.9). With this construction, we can find a wave with a wave vector lying inside the Brillouin zone that is equivalent to any given wave in the whole of $\boldsymbol{k}$-space.

The Brillouin zone is constructed as follows:
Let the origin of the reciprocal lattice be O and let M be one of its nodes. The median plane of the segment $[O, M]$ divides space into two parts: the one containing O contains all the points closer to O than M. Median planes are similarly constructed for all the nodes. The whole set of these planes forms a closed polyhedron around the origin and centred on the origin: this is the Brillouin zone.

In Fig. 1.10, we show as examples the Brillouin zones of a two-dimensional lattice and of the reciprocal lattice of a face-centred cubic lattice.

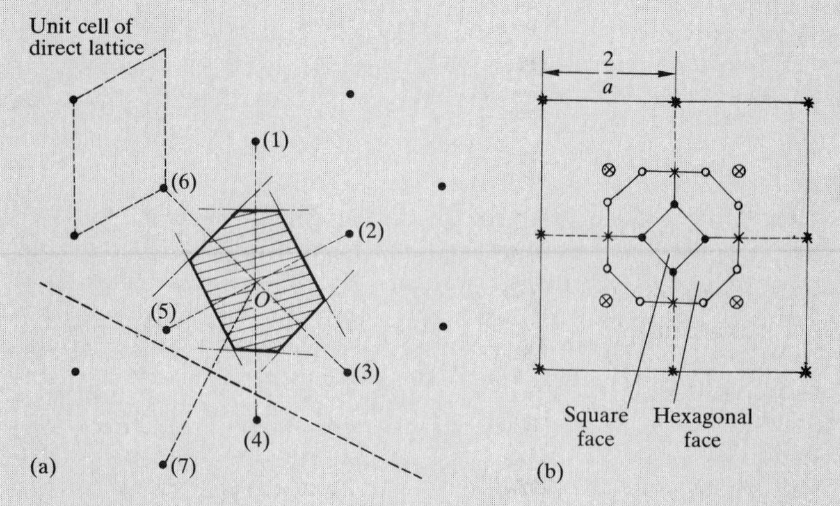

Fig. 1.10. (a) Brillouin zone of a plane lattice. It is defined by the perpendicular bisectors of the vector joining the origin to the nodes 1, 2, 3, 4, 5, and 6. For any other node, the perpendicular bisector lies outside the Brillouin zone. (b) The reciprocal lattice (a body-centred cubic lattice of unit cell parameter $2/a$) of a face-centred cubic direct lattice of unit cell parameter a. The symbols indicates various heights perpendicular to the plane of the diagram: nodes at heights of 0, $2/a$, . . . are indicated by * and those at $\pm 1/a$, . . . by ⊗. The Brillouin zone has the following vertices: 8 at height 0 (○), 8 at heights $\pm 1/a$ (●), and 8 at heights $\pm 1/2a$ (×); and has 8 hexagonal faces and 6 square faces. All its edges are of length $1/a\sqrt{2}$.

3 *Frequencies of the modes*

There are three frequencies corresponding to the three modes associated with each of the N wave vectors. In a crystal, the frequencies of the two transverse modes may in practice be the same, as it would be in an isotropic solid.

For small wave vectors (i.e. long wavelengths, say above $1\,\mu m$) the situation ties in with the case of 'macroscopic' waves in the continuous solid: the discontinuous atomic structure does not have any effect because of the large number of atoms on the scale of a wavelength. Under these conditions, the mode frequency is simply proportional to the wave vector : $\omega = \upsilon k$, where υ is the velocity of wave propagation in the medium.

There are three velocities: one for longitudinal waves giving the speed of sound; the other two, which may coincide, for transverse waves. Crystal anisotropy causes the velocity of propagation to vary as the direction of the wave vector changes in relation to the crystal axes.

When the wave vector is no longer small in comparison with its limiting value, **dispersion** occurs: in other words, as in the linear chain, ω/k is not constant. This is caused by the atomic discontinuity in the crystal. For each direction of wave propagation and for each type of wave, transverse or longitudinal, there is a maximum value of the frequency beyond which a wave cannot be propagated in the crystal.

4 *Energy of crystal vibrations*

The Einstein model replaced the vibrating crystal with a set of $3N$ independent linear oscillators having the same frequency. When atomic interactions are taken into account, we are led to analyse thermal agitation in a different way: the 'element' becomes a normal mode of vibration of the crystal involving all its atoms. The various modes are independent and all the modes are superposed in order to describe the atomic agitation in a crystal.

There are $3N$ modes for N atoms. We see that, in spite of the differences between the two models, they both have the same number of 'elements', $3N$. In the classical theory, there is equipartition of energy between the normal modes, each having an amount of energy $k_B T$. The total molar energy is therefore the same as in the Einstein model. *At high temperatures, where classical theory is valid, Dulong and Petit's law is once again satisfied.*

At low temperatures, however, the behaviour of the heat capacity proves that the energy of the vibrational modes in the crystal is quantized, as is the vibration of the Einstein oscillator. The quantum of energy of the mode with frequency ω, known as a **phonon**, is $\hbar\omega$. At equilibrium, according to Bose–Einstein statistics (p. 10), the mean number of phonons in this mode at temperature T is:

$$\bar{n} = \frac{1}{\exp\left(\frac{\hbar\,\omega}{k_B T}\right) - 1}$$

and the mean energy per mode is therefore:

$$\bar{E} = \frac{\hbar\,\omega}{2} + \frac{\hbar\,\omega}{\exp\left(\frac{\hbar\,\omega}{k_B T}\right) - 1}\,.$$

The difference between this model and that of Einstein is that there is no longer only a single vibration frequency ω_E: the phonon frequency, depending on the mode, varies from 0 to ω_{max}.

Finding the energy of a mode of wave vector k requires a calculation of its frequency. This is a very difficult operation: it has to be carried out afresh for each crystal and it involves parameters defining the atomic interactions, followed by a summation of the contributions from all the modes. To simplify the calculation, we make use of an approximation suggested by Debye.

Debye's model

First of all, it is assumed that the distribution of the modes is spherically symmetrical, i.e. that the frequency of a mode depends only on the magnitude of its wave vector. This is quite a good approximation for crystals of high symmetry, such as cubic crystals.

Secondly, dispersion is neglected: it is assumed, just as in the case of small wave vectors, that the frequency is proportional to the wave vector: $\omega = \upsilon k$, where υ, which has the dimensions of a velocity, is chosen to lie between the velocities of the longitudinal and transverse vibrations.

We show in Box 5 how the number of modes with frequencies lying between ω and $\omega + d\omega$ can be calculated, and also how the total number of modes with frequencies lying below a given frequency can be obtained by making some approximations. However, since the dispersion curve is no longer involved, the upper limit on ω has disappeared. Nevertheless, we know that the number of modes is finite and equal to $3N$. Debye therefore introduced a cut-off at a maximum frequency ω_D, so chosen that the total number of modes had the correct value of $3N$. The value of ω_D is expressed in terms of a temperature, Θ_D, known as the Debye temperature, such that the energy $\hbar\omega_D$ of the mode of frequency ω_D is equal to the classical energy at the temperature Θ_D, and thus that

$$\Theta_D = \frac{\hbar}{k_B}\omega_D$$

The total vibrational energy of the crystal at the temperature T can be calculated with the Debye approximation in terms of the single parameter T/Θ_D.

Box 5

Spectral density of low frequency vibrational modes. The heat capacity in Debye's model

Consider the possible modes of vibration of a volume in the shape of a cube with fixed faces defined by the coordinates $x = 0$, $x = L$; $y = 0$, $y = L$; $z = 0$, $z = L$. In general, any wave with a given polarization corresponding to a possible vibration can be represented by the real part of the complex number:

$$A \exp i[\omega t - (k_x x + k_y y + k_z z)].$$

The conditions for reflection and zero amplitude over the face $x = 0$ mean that this wave must combine with the reflected wave of wave vector $(-k_x, k_y, k_z)$ in such a way that the two cancel each other out at $x = 0$. The expression for the resultant wave can then be written:

$$A \sin(k_x x) \exp i[\omega t - (k_y y + k_z z)].$$

The same argument applied to the faces $y = 0$ and $z = 0$ leads to the final resultant wave:

$$A \sin(k_x x) \sin(k_y y) \sin(k_z z) \exp i\omega t.$$

Note that definite signs must be allocated to k_x, k_y, k_z: for example, they may be taken as *positive* numbers. This is because the k_x wave is combined with the $-k_x$ wave so that the two are no longer independent, and similarly with the k_y and k_z waves.

We now turn to the conditions for reflection and zero amplitude over the faces $x = L$, $y = L$, $z = L$. These conditions can be written:

$$\sin(k_x L) = 0, \text{ i.e. } k_x = n_x\pi/L$$
$$\sin(k_y L) = 0, \text{ i.e. } k_y = n_y\pi/L$$
$$\sin(k_z L) = 0, \text{ i.e. } k_z = n_z\pi/L$$

where n_x, n_y, n_z are *positive* integers ($n_x = 0, 1, 2 \ldots$).
There is therefore one point **k** *per elementary cube of side* π/L *in* **k** *space.*

We now ask the question: how many points k are there in k space giving rise to independent waves having the moduli of their wave vectors lying between k and $k + \Delta k$? Since k_x, k_y, and k_z are positive numbers, the points $\boldsymbol{k}$ lie in an octant of the volume between the two spheres of radii k and $k + \Delta k$, at a density of one per elementary cube of side π/L. The number of points $\boldsymbol{k}$ is therefore:

$$\frac{1}{8} \cdot \frac{4\pi k^2 \Delta k}{\left(\frac{\pi}{L}\right)^3} = \left(\frac{V}{2\pi^2}\right) k^2 \Delta k$$

where V denotes the total volume, L^3. However, we have to deal with vibrational waves in a solid, in which three independent modes (one longitudinal and two transverse) are associated with each vector $\boldsymbol{k}$. The number of modes with wave vectors lying between k and $k + \Delta k$ is therefore:

$$\frac{3V}{2\pi^2} k^2 \Delta k.$$

For such waves, we know that the relationship between ω and k, the dispersion relation, becomes linear at low frequencies, i.e. $\omega = vk$. In this expression, the propagation velocity v is assumed to be the same for transverse and longitudinal waves. The number of independent modes whose frequency lies between ω and $\omega + \Delta\omega$ is therefore:

$$f(\omega)\Delta\omega = \frac{3V}{2\pi^2 v^3} \omega^2 \Delta\omega.$$

The function f, known as the **spectral density**, is thus proportional to ω^2 at low frequencies.

Debye made the following approximation: this ω^2 law remains valid up to a maximum frequency ω_D, called the **Debye frequency,** calculated in such a way that the total number of modes (i.e. the integral of f over the interval $[0, \omega_D]$) is equal to $3N$. We therefore have that

$$\frac{3V}{2\pi^2} \frac{1}{v^3} \int_0^{\omega_D} \omega^2 d\omega = 3N$$

i.e.

$$\frac{V}{2\pi^2}\left(\frac{\omega_D}{\upsilon}\right)^3 = 3N.$$

This relationship connects the cut-off frequency ω_D with the speed of sound:

$$\omega_D = (6\pi^2 n)^{1/3}\upsilon$$

where $n = N/V$ is the number of atoms per unit volume.

Using ω_D, the density of states referred to a mole can then be written in the Debye approximation in the form:

$$f(\omega) = 9N_A\omega^2/\omega_D{}^3.$$

The calculation of the heat capacity is then continued by generalizing Einstein's calculation. In a frequency interval between ω and $\omega + \Delta\omega$ there are $f(\omega)\Delta\omega$ modes or oscillators with a mean energy at temperature T given by $[\bar{n}(\omega) + \frac{1}{2}]\hbar\omega$, where $\bar{n}(\omega)$ is the mean quantum number of an oscillator at temperature T, given by the Bose–Einstein formula (Box 2).

These modes thus contribute the following energy to the system:

$$\Delta E = \left[\bar{n}(\omega) + \frac{1}{2}\right]\hbar\,\omega\, f(\omega)\,\Delta\omega.$$

The total energy is obtained by summing these contributions over the whole frequency 'band' from $\omega = 0$ to $\omega = \omega_D$:

$$E = \int_0^{\omega_D}\left[\bar{n}(\omega) + \frac{1}{2}\right]\hbar\,\omega\, f(\omega)\,d\omega.$$

The heat capacity is then obtained by differentiating this expression with respect to temperature (which only appears in $\bar{n}(\omega)$):

$$C = k_B\int_0^{\omega_D}\left(\frac{\hbar\,\omega}{k_BT}\right)^2\frac{\exp\dfrac{\hbar\,\omega}{k_BT}}{\left(\exp\dfrac{\hbar\,\omega}{k_BT} - 1\right)^2} f(\omega)d\omega.$$

Note that this formula is a priori valid for all forms of the density of states, but only the Debye approximation enables the calculation to be followed through in a simple way. By putting $x = \hbar\omega/k_B T$, we obtain:

$$C = 9R\left(\frac{T}{\Theta_D}\right)^3 \int_0^{\frac{\Theta_D}{T}} \frac{e^x}{(e^x - 1)^2} x^4 dx.$$

In this formula, we have introduced the Debye temperature, $\Theta_D = \hbar\omega_D/k_B$ (p. 25). It can be seen that C is a function of the single parameter T/Θ_D.

It should be pointed out that, when T is very large, the integral involves only small values of x and this enables $e^x/(e^x - 1)^2$ to be replaced by $1/x^2$. We then find once again that:

$$C = 9R\left(\frac{T}{\Theta_D}\right)^3 \int_0^{\frac{\Theta_D}{T}} x^2 dx = 3R$$

which is none other than Dulong and Petit's law. On the other hand, when T is close to 0 K, we use the expression for the integral:

$$\int_0^{+\infty} \frac{e^x}{(e^x - 1)^2} x^4 dx = \frac{4\pi^4}{15}$$

to find that:

$$C = \frac{12\pi^4}{5} R\left(\frac{T}{\Theta_D}\right)^3.$$

When T tends to zero, the heat capacity tends to zero like T^3, i.e. more slowly than the exponential variation given by the Einstein calculation (Fig. 1.3).

It is found that the Debye formula, which is an interpolation between two regions (low and high temperatures) where it is strictly true, gives a remarkably good approximation to the variation of heat capacity with temperature for a large number of substances. In fact, a completely rigorous calculation would involve knowing the exact

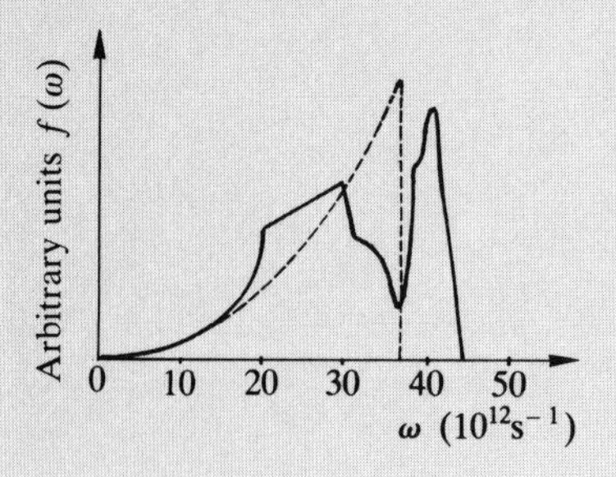

Fig. 1.11. Spectral density of the vibrational modes in copper: (______) theoretical curve taking into account the crystal dynamics of copper, (----) Debye approximation.

form of the density $f(\omega)$ of the modes of vibration in each case. Even if it were true that $f(\omega)$ always varied as ω^2 at sufficiently low frequencies, its behaviour at higher frequencies depends on subtleties in the dynamics of atomic vibrations that are specific to each crystal. In particular, crystal anisotropy produces a difference between the bands of transverse and longitudinal modes, which do not a priori have the same form. Figure 1.11 shows the spectral distribution for copper established from a fairly refined dynamic model, and how it compares with the Debye approximation.

Comparison of the Debye model with experiment

We have seen that the Einstein model already gives good agreement with the experiment values of the heat capacity. In the high temperature region, the Debye and Einstein models are both satisfactory, since they tend towards the elementary classical model leading to Dulong and Petit's law. The replacement of a single frequency by a rather narrow band of frequencies does not have any significant consequences. For a given solid, it is possible to find a Debye temperature which accounts quite well for the experimental results (Table 1.2).

At very low temperatures, on the other hand, the Debye model is better. This is because it includes low frequency modes which are capable of being excited, whereas the oscillators at the single Einstein frequency are not excited. The heat capacity predicted by Debye is greater than that predicted by Einstein and is in agreement with experiment (Fig. 1.3).

The temperatures given in tables of constants are quite scattered, depending on whether Θ_D is chosen to give better agreement with heat capacities at high or low temperatures. Debye temperatures can also be derived from mechanical properties.

Table 1.2 Some Einstein and Debye temperatures

Element	Θ_E(K)	Θ_D(K)
C (diamond)	1500	3000
Al	300	390
Cu	250	320
Mo	290	380
Ag	165	226
Pb	70	90

Contributions made by theories of the heat capacity of monatomic solids

In our search for a theoretical explanation of the heat capacity of a crystal, we have introduced several concepts which turn out to have a wider significance than appears at first sight.

In the first approximation (Dulong and Petit's law), the internal thermal energy of a solid depends only on the number of atoms it contains; it is independent both of the nature of the atom and the crystal structure. The atom is subjected to the action of all its neighbours, so that its motion boils down to vibration about a fixed point. In spite of the great complexity of this situation, the thermal energy of an atom in the solid is simply twice the mean kinetic energy of an atom moving freely in a perfect gas at the same temperature: $3k_B T$ instead of $3k_B T/2$.

This high temperature limit for the heat capacity per atom is a basic result that is still yielded by the most refined of the theories which have replaced the simplest form of classical theory.†

The second new concept, the **quantization of energy**, was introduced to explain the behaviour of the heat capacity at very low temperatures. The existence of the photon, the energy quantum of electromagnetic radiation, had already been demonstrated by Planck in connection with black body radiation. Einstein quantized elastic vibrations with the **phonon**.

The third new concept to be introduced was not brought in merely to improve agreement with experiment, but was proposed mainly because the Einstein model, with its independent atomic oscillators, was artificial. It seemed difficult to ignore the strong bonds between neighbouring atoms which are the very reason for the existence of solids.

The basic phenomena were assumed to be the **modes of vibration**, which involve the crystal as a whole and which are also quantized: the thermal agitation of the atoms is then the result of superposing all the possible modes of vibration, which can be obtained accurately from the theory. Although this model is very different from that of Einstein, it yields more or less the same results. In addition, however, the agreement with experimental measurements at very low temperatures is significantly improved.

The refinements of the Debye theory thus contribute nothing conclusive to the problem of the heat capacity. Nevertheless, it does bring the phenomenon of crystal vibrations into the picture, and the importance of this is much wider than the single question of the heat capacity. It is true that the experimental study of heat capacity provides no direct and

†The experimental values of C may be greater than the theoretical value of $3R$. This is an effect arising from the anharmonicity of the vibrations, from the contribution of the free electrons in the case of metals and from the formation of vacancies in the lattice.

accurate information about the elastic waves in the crystal, but fortunately there are other physical methods for revealing them and providing more details about their properties. This is the aspect we are now going to investigate. It is important to do this because, in any model of the atomic structure of solids, thermal agitation is an essential complement to the static and perfectly regular picture that springs from crystallographic studies: we are approaching the field of crystal dynamics.

Inelastic scattering of neutrons by phonons

Vibrational waves are excited spontaneously in a crystal by the thermal agitation of atoms. Their amplitude increases as the temperature rises. These elastic waves are often called **phonons**, but we should point out that the term is used in two different senses. Its other, and principal, meaning is that of the quantum of vibrational energy $\hbar\omega$ for a frequency ω.

Phonons which can be propagated through a crystal have a minimum wavelength of the order of twice the linear dimension of the crystal unit cell and a maximum frequency of around 10^{12}Hz or 1 terahertz (THz). This corresponds to a velocity of propagation for elastic waves of the order of $5000\,\mathrm{m\,s^{-1}}$.

Thermal phonons are of the same type as those of the elastic waves excited in solids by an external mechanical or acoustic source, but the latter have much lower frequencies. For example, the frequency of ultrasonic waves has a practical upper limit of 10^9 Hz, which is very low in comparison with the 10^{13} Hz typical of thermal phonons.

The most fruitful technique for studying phonons uses **inelastic neutron scattering**. This enables the phonon frequency to be measured as a function of its wave vector, i.e. it enables the dispersion curve to be determined. However, the technique is complex and tricky and no details will be given: we shall merely describe the information that can be extracted from the method by giving a few examples. First of all, however, we look at a closely related phenomenon: that of X-ray scattering.

Consider a crystal *without defects* but traversed by the set of waves due to thermal agitation. The crystal is then irradiated by a monochromatic X-ray beam which gives rise to secondary waves through scattering by each atom. When they emerge from the crystal, the secondary wavelets scattered by all the atoms interfere, and the complex result of such interference is recorded.

In the first place, by far the most important part of what is observed is the **diffraction pattern**: this is the set of **Bragg reflections** (SM, p. 71) used for the determination of crystal structures. When the geometry of the experimental arrangement (the directions of the primary and secondary rays in relation to those of the crystal axes, the primary wavelength) is such that there is no Bragg reflection, the interference that occurs would completely cancel out any resultant X-ray scattering if the atoms in the crystal were stationary.

If any scattering is observed outside the Bragg reflections, it will be due to the vibrations of the atoms around their theoretical positions. The remarkable fact that makes this inelastic scattering a useful method is that the only atomic waves involved in it are those with wave vectors determined by the geometry of the experiment: all the others, infinitely more numerous, have no effect in the given set of observational conditions. It is this property that enables the scattering experiment to provide a natural analysis of the waves due to thermal agitation in terms of their wave vectors, as we have done theoretically.

Consider an active elastic wave, i.e. one that causes the scattering observed in a certain direction. What information will the scattered radiation yield? Calculation of the interference produced by the crystal traversed by this single wave shows that the scattered radiation no longer has the same frequency as the incident radiation, Ω, but a modified frequency $\Omega' = \Omega \pm \omega$, where ω is the frequency of the elastic wave causing the scattering. For X-rays Ω and ω are of the order of 10^{18} and 10^{12} Hz respectively, and so the theoretically predicted frequency shift is too small to be detected experimentally.

The situation is different for neutrons. The formula for the frequency shift, obtained by calculation of the interference that occurs, can be written (in the special language of the method) in the form: $\hbar\Omega' = \hbar\Omega \pm \hbar\omega$, and interpreted in terms of particles. $\hbar\Omega'$ and $\hbar\Omega$ are the photons of the incident and scattered radiation respectively, or the energies of the incident and scattered neutrons, while $\hbar\omega$ is the phonon energy for the elastic wave.

In this type of scattering process, described as **inelastic**, the incident neutron interacts with the wave due to the atomic vibrations, so that the scattered neutron has an energy that is greater than or less than that of the incident neutron: what was given to or acquired by the elastic wave can only be a phonon $\hbar\omega$. This is what is described by the formula for the frequency shift.

The difference between the technique using X-rays and that using neutrons lies in the orders of magnitude. The neutrons used in these experiments, termed thermal neutrons, have an energy of a few meV, i.e. comparable with that of the phonon. The energy of the X-ray photon, on the other hand, is of the order of at least several keV, thus confirming that the energy change of the X-ray photon is too small to be detected.

The difference between the energies of the incident and scattered neutrons is easily detectable and, since it corresponds to the energy of the phonon responsible for the scattering, the frequency of the elastic wave can thus be measured. Moreover, since the geometry of the experiment determines the value of the wave vector, *inelastic neutron scattering enables the dispersion curve for the elastic waves in the crystal to be determined.*

Whereas X-ray or neutron **diffraction** reveals the mean positions of the atoms in the crystal, inelastic **scattering** involves crystal dynamics: the dispersion curve for phonons is determined by interatomic forces and, conversely, it is theoretically possible to find out about these forces from the curve.

Figure 1.12 shows dispersion curves for a copper crystal determined by inelastic neutron scattering. The wave vector is parallel either to a side of the cubic unit cell or to a diagonal on one of its faces; it varies from 0 to the limit of the Brillouin zone. A longitudinal wave and transverse waves occur, the latter being coincident in the first case. The slope of the tangent at the origin gives the wave propagation velocities (acoustic or elastic) which can be measured directly. The agreement with the neutron scattering experiments is very satisfactory.

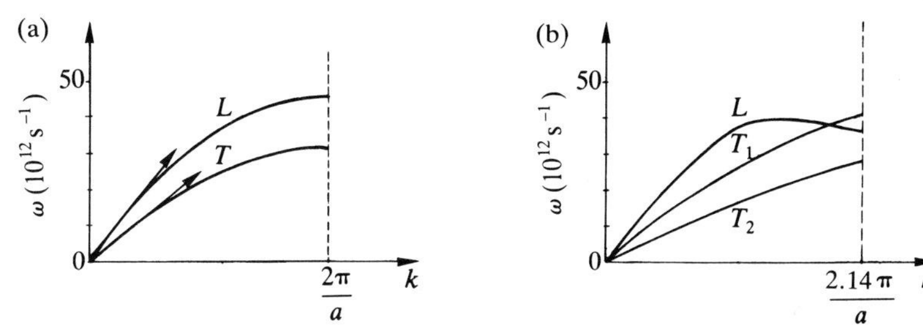

Fig. 1.12. Dispersion curves for the phonons in a copper crystal (face-centred cubic lattice with unit cell parameter $a = 0.361\,\text{nm}$) with the wave vector k (a) along a cube edge and (b) along a side face diagonal. Tangents at the origin are calculated from the elastic constants of the crystal (By kind permission of B. N. Brockhouse).

The heat capacity of polyatomic crystals

In a monatomic crystal, all the atoms play exactly the same role: the theoretical consequence of this, verified at high temperatures, is that the mean vibrational energy is independent of the nature of the atom and has the value of $3k_BT$ per atom. This result could be extended to a polyatomic crystal if it is assumed, in spite of their different bonds, that all the atoms in the compound vibrate independently. If p is the number of atoms per molecule, a mole contains N_Ap atoms. The molar thermal energy is then $N_Ap.3k_BT$ and the molar heat capacity is $3Rp$.

It is observed that the heat capacity of a compound decreases towards low temperatures and becomes zero at 0 K. As T increases, it tends to a limit which should be compared with the above predicted value. Table 1.3

gives data for compounds with different numbers of atoms per molecule.

It can be seen that there is often quite good agreement (to within 10 per cent) between the data and the simplistic theory. When there is a large discrepancy, the experimental value is less than the theoretical value, as it is in the exceptional monatomic cases (diamond, p. 7).

To proceed any further, the elastic modes of vibration of the crystal must be described. The complication here is that all the displacements of all the different atoms cannot be represented by a single wave: not only are the displacements of the whole atomic motif being propagated from one unit cell to another, but superimposed on these are the deformations of the motif itself due to the motion of individual atoms relative to their neighbours.

We shall take a simple example, which has the advantage that it can be dealt with using simple calculations and that certain conclusions which can be drawn from it are capable of being generalized to a crystal. The example we choose is that of a linear chain of two atomic species of different masses M_1 and M_2, alternating regularly at the nodes of a one-dimensional lattice of period $2a$. As in the case of a set of identical atoms, we restrict our treatment to longitudinal vibrations. The interaction between two neighbours (here all the pairs are of the same nature) is represented by a force proportional to the difference between the real distance and the equilibrium distance a. Box 6 gives the detailed calculations for waves being propagated along the line of atoms.

Table 1.3 Molar heat capacities of polyatomic crystals

Compound	Theoretical value ($3R \times p$)	Experimental value ($J\,K^{-1}\,mol^{-1}$)	Ratio
NaF		47	0.94
NaCl		55	1.10
NaBr		52	1.04
NaI		54	1.08
KF	$3R \times 2 = 50$	48	0.96
KCl		54	1.08
KBr		51	1.02
KI		52	1.04
MgO		44	0.88
CaO		53	1.06
CaF_2	$3R \times 3 = 75$	66	0.85
SiO_2		73	0.97
H_2O (ice)		37.5	0.5
$BaSO_4$	$3R \times 6 = 150$	128	0.85

Heat capacities of liquid compounds

C_6H_6	$p = 12$ $25 \times 12 = 300$	133	0.44
C_2H_6O	$p = 9$ $25 \times 9 = 225$	115	0.51
$C_{30}H_{62}$	$p = 92$ $25 \times 92 = 2300$	1224	0.53
H_2O	$p = 3$ $25 \times 3 = 75$	75	1.0
CS_2	$p = 3$ $25 \times 3 = 75$	76	1.0
CCl_4	$p = 5$ $25 \times 5 = 125$	119	0.95
CH_3CO_2H $(C_2O_2H_4)$	$p = 8$ $25 \times 8 = 200$	118	0.59
CH_3COCH_3 (C_3H_6O)	$p = 10$ $25 \times 10 = 250$	128	0.53
$C_4H_{10}O$	$p = 15$ $25 \times 15 = 375$	170	0.45

Box 6

Linear chain of two different atoms

Two different kinds of atom of masses M_1 and M_2 are located alternately along a straight line and are separated when at rest by a distance a (Fig. 1.13). All the nearest neighbours pairs are identical and consist of two dissimilar atoms. In all the pairs, the force experienced by any atom is proportional to the distance between the two atoms, with a single constant of proportionality K.

The masses M_1, of even order $(2n)$, whose equation of motion can be written

$$M_1 \frac{d^2 u_{2n}}{dt^2} = -K(2u_{2n} - u_{2n+1} - u_{2n-1}),$$

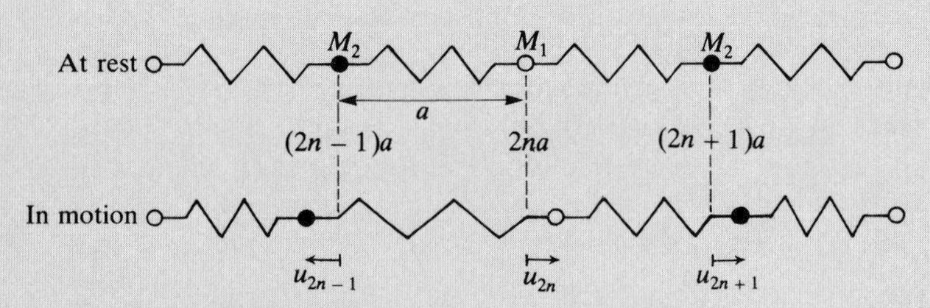

Fig. 1.13. A linear chain with two different kinds of atom.

must be distinguished from the masses M_2, of odd order $(2n+1)$, whose equation of motion is:

$$M_2 \frac{d^2 u_{2n+1}}{dt^2} = -K(2u_{2n+1} - u_{2n} - u_{2n+2}).$$

Just as in the case of the monatomic chain, we shall seek solutions in the form of waves but this time the amplitude of the wave will be different for the masses M_1 and M_2. Let these amplitudes be U_1 and U_2. We shall put:

$$u_{2n} = U_1 \cos[\omega t - 2nka]$$
$$u_{2n+1} = U_2 \cos[\omega t - (2n+1)ka].$$

By substituting these expressions into the equations of motion, we obtain a system of equations that is linear and homogeneous in U_1 and U_2:

$$\begin{cases} (2K - M_1\omega^2)\,U_1 - 2K\cos ka U_2 = 0 \\ -2K\cos ka U_1 + (2K - M_2\omega^2)U_2 = 0. \end{cases}$$

Such a system of equations only has non-zero solutions for U_1 and U_2 if the determinant of its coefficients is zero. By equating this determinant to zero, we obtain:

$$M_1 M_2 \omega^4 - 2K(M_1 + M_2)\omega^2 + 4K^2 \sin^2 ka = 0$$

which is a quadratic equation in ω^2. The two solutions for ω^2, when plotted as functions of k, yield curves known as the **acoustic branch** and the **optical branch**, for reasons explained later.

Acoustic branch (Fig. 1.14)

$$\omega_a^2 = K\left(\frac{M_1 + M_2}{M_1 M_2} - \frac{\sqrt{(M_1 - M_2)^2 + 4M_1M_2\cos^2 ka}}{M_1 M_2}\right).$$

This solution tends linearly to zero as k tends to zero (long

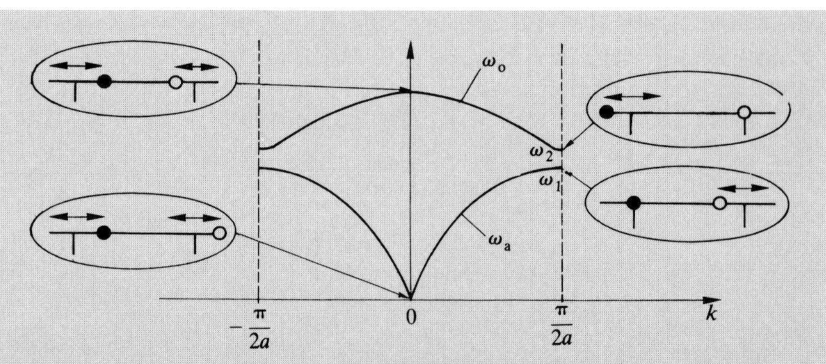

Fig. 1.14. Dispersion curves for the diatomic chain (masses M_1 [○] and M_2 [●]). The small sketches show how the atoms vibrate at $k = 0$ and $k = \pi/2a$ for the optical and acoustic branches.

wavelengths). This can be seen by putting $\cos ka = 1 - k^2a^2/2$ and simplifying, when we obtain, for small ka:

$$\omega_a \approx a \sqrt{\frac{2K}{M_1 + M_2}}\,|k|.$$

When $M_1 = M_2$, the expression for the speed of sound obtained in the case of a linear monatomic chain is obtained again. Moreover, when k tends to zero, the amplitudes U_1 and U_2 become equal: atoms (1) and (2) vibrate in phase.

Optical branch (Fig. 1.14)

$$\omega_0^2 = K\left(\frac{M_1 + M_2}{M_1M_2} + \frac{\sqrt{(M_1 - M_2)^2 + 4\,M_1M_2\cos^2 ka}}{M_1M_2}\right).$$

This solution tends to a non-zero limiting value as k tends to zero:

$$\omega_0\,(k = 0) = \sqrt{\frac{2K(M_1 + M_2)}{M_1M_2}}\,.$$

Close to this limit, we find that

$$U_1/U_2 = -\,M_2/M_1$$

so that atoms (1) and (2) vibrate in phase opposition.

Limit when M_1 tends to M_2

It is interesting to see how the solution for the linear monatomic chain is recovered as M_1 tends to M_2. As long as M_1 remains different from M_2, there is a non-zero difference (a *gap*) between the two branches at $k = \pm\pi/2a$, where $\omega_a = \sqrt{2K/M_1}$ and $\omega_0 = \sqrt{2K/M_2}$. The acoustic and optical branches tend quadratically to these limiting values as the quantity $(\pi/2a - k)$ tends to zero:

$$\omega_a^2 \approx \frac{2K}{M_1} - A_1\left(\frac{\pi}{2a} - k\right)^2 + \ldots$$

$$\omega_0^2 \approx \frac{2K}{M_2} - A_2\left(\frac{\pi}{2a} - k\right)^2 + \ldots.$$

When M_1 tends to M_2, not only does the difference between ω_a and ω_0 tend to zero but both the coefficients A_1 and A_2 tend to infinity like $1/(\pi/2a - k)$ so that, when $M_1 = M_2$, we revert to a linear variation at $k = \pi/2a$ as in Fig. 1.6.

Another very subtle point is the following: when the condition $M_1 = M_2$ is achieved by passing to the limit $(M_1 \to M_2)$, the period of the system in real space remains constant at $2a$, whereas when M_1 actually equals M_2 it becomes a. We thus revert to the curve of the monatomic linear chain (which is normally defined over the complete

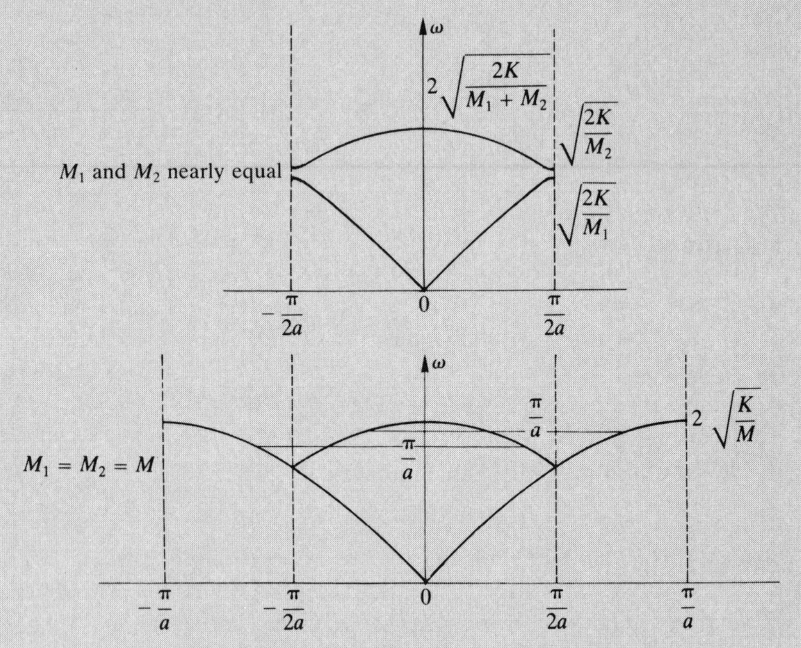

Fig. 1.15. Passing to the limit $M_1 = M_2$ (from the diatomic to the monatomic chain).

Brillouin zone $\pm\pi/a$) restricted to the half Brillouin zone $\pm\pi/2a$: the sections for $|k| > \pi/2a$ have been shifted by a reciprocal lattice vector (Fig. 1.15).

Elastic waves in polyatomic crystals

We now comment on the results of the calculation for the linear chain and generalize them to the three-dimensional crystal.

1. The wavelength of the vibration has a minimum value which is twice the period of the linear lattice, i.e. it is $2a$. As in the case of the chain of identical atoms, all the waves can be represented by restricting the wave vector to the interval $-\pi/2a < k \leqslant \pi/2a$. In a crystal, the wave vector lies inside the Brillouin zone of the reciprocal lattice (see Box 4).

Furthermore, for a crystal of finite extent, the ends of the wave vectors have a regular distribution depending on crystal size. The consequence of this is that the number of wave vectors is finite and *equal to the number of unit cells*, as in the monatomic crystal.

2. The particular feature introduced by the different types of atom is that, corresponding to any wave vector k, *there are two distinct types of wave* having different frequencies, ω_a and ω_0, with $\omega_a < \omega_0$.

For the lower frequency, the dispersion curve $\omega_a = f(k)$ (Fig. 1.14) is similar to that of the simple lattice: ω_a is zero when $k = 0$. For small k, ω_a is proportional to k and the amplitudes for atoms (1) and (2) are more or less equal. This means that, for sufficiently long wavelengths, two neighbouring atoms vibrate together without an appreciable variation in their distance apart: we can say roughly that it is the whole unit cell which is oscillating and it is the motion of the unit cell which is being propagated. These are **acoustic waves:** they are an extension, on the short wavelength side, of the elastic waves excited by an external acoustic or mechanical source and they propagate at the speed of sound. This is the essential feature of the acoustic mode. However, when k increases (i.e. when k is no longer small compared with k_{max}), the acoustic mode is complicated because the unit cell becomes more and more deformed. Thus, when we reach k_{max}, all the heavier atoms are vibrating in phase and the intermediate lighter atoms remain stationary (Fig. 1.14).

The other type of wave has a frequency ω_0 which is a maximum when $k = 0$. As k increases, ω_0 slowly decreases and reaches a minimum at k_{max}. At the limiting value $k = 0$, the two atoms in the unit cell are vibrating around their centre of mass, which remains fixed. Each set of atoms of the same type vibrate in phase and the two sets vibrate with opposite phases. There is no propagation and no overall displacement of the unit cell, merely a periodic deformation. The vibrations of this type in a row of atoms are similar to those of a molecule responsible for its optical properties (the absorption and emission of light). That is why this type of wave is called the **optical mode.**

For non-zero values of k, and increasingly as k increases, the vibration becomes a mixture of displacement and deformation of the unit cell. Thus, for $k = k_{max}$, the heavier atoms remain stationary and the others vibrate in phase.

To sum up, the dispersion curve $\omega = f(k)$ for the diatomic row consists of two distinct branches, the acoustic and the optical. In the first of these, the frequency varies from zero to a maximum ω_1 at $k = \pm \pi/2a$; in the second, the frequency decreases from ω_0 to a minimum ω_2 at $k = \pm \pi/2a$. Suppose that it is possible to excite a vibration of frequency ω using an external source. The row of atoms can be set into vibration provided that ω lies in one of the two bands $[0, \omega_1]$ and $[\omega_2, \omega_0]$. Outside these 'allowed' frequencies, there is a forbidden band or *gap* from ω_1 to ω_2. This constitutes a frequency filter and is a direct result of the periodicity of the structure. We shall meet another example of this when dealing with electrons in crystals (p. 90).

3. As in the case of the monatomic crystal, the set of modes of vibration of a polyatomic crystal corresponds to wave vectors whose ends are distributed with a uniform density inside the Brillouin zone. There are N wave vectors if the crystal consists of N unit cells.

In a diatomic crystal, just as in the case of a monatomic crystal, a given wave vector must be considered as corresponding to three associated waves having different vector amplitudes: one is a longitudinal wave and the other two are transverse. If the frequencies of the latter are equal, a crystal with two atoms per unit cell has two dispersion curves for the acoustic modes and two for the optical modes. A mole of the crystal contains N_A unit cells and thus $2N_A$ atoms, so that there are $3N_A$ acoustic modes and $3N_A$ optical modes, which is $3N_A \times 2$ in all, i.e. three times the total number of atoms, as in the case of a monatomic crystal.

This can be generalized to a crystal containing p atoms per unit cell. The total number of modes is $3N_Ap$, and of these $3N_A$ are acoustic modes in which the unit cell vibrates *en bloc*, and $3N_A(p - 1)$ are optical modes corresponding to different deformations of the unit cell. At high temperatures, where the classical theory is valid, each mode has an energy k_BT and we revert to the value $3Rp$ for the molar heat capacity already predicted (p. 33).

4. The modes of vibration of a crystal can be classified according to their frequency. The number of discrete modes with frequencies lying between ω and $\omega + \Delta\omega$ can be found as a function of ω, and this yields a curve showing the density of modes or the spectral density (Fig. 1.16(b)). The

calculation involves crystal dynamics and is difficult because the interatomic forces are not known with sufficient accuracy. From this curve, and from the formula for the mean energy per mode of frequency ω as a function of temperature as given by the Bose–Einstein equation (p.11), the heat capacity of a polyatomic crystal can be calculated. At temperatures that are not too high, particularly at ordinary temperatures, many of the optical modes distributed in high frequency bands are not excited. The heat capacity is then lower than the Dulong and Petit value: this is what can be observed in the 'exceptional' cases of Table 1.3.

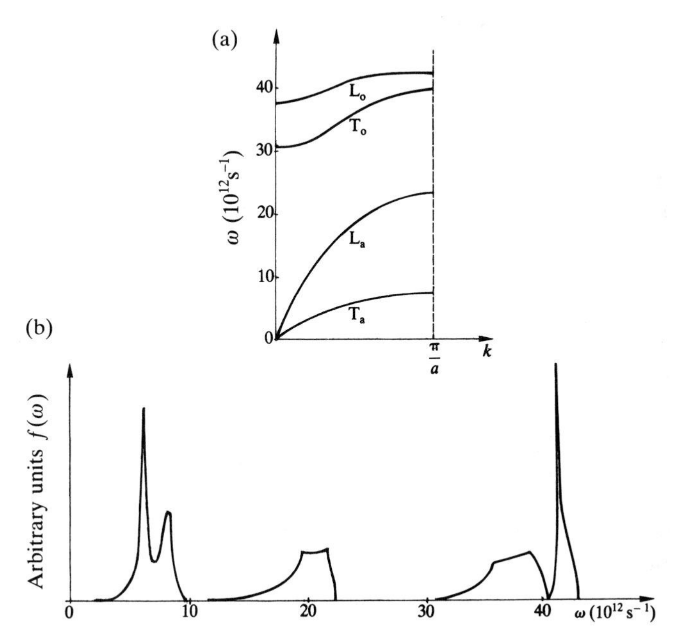

Fig. 1.16. Experimental dispersion curve for phonons in a polyatomic crystal (CuCl). (a) Wave vector along the edge of the cubic unit cell; (b) spectral density of phonons in the CuCl crystal (by kind permission of B. Hennion).

5. Experimental data on waves in polyatomic crystals are provided by inelastic neutron scattering, since the frequency of a wave with a given wave vector can be measured. With polyatomic crystals, there are several frequencies per wave vector, thus confirming the multiplicity of branches of the dispersion curve. Since the intensity scattered by a mode depends on the orientation of the vector amplitude of the wave, longitudinal and transverse waves can be distinguished from each other. However, the resolution of the equipment used to measure the frequency is not very high and the results are not easy to interpret unless the branches of the dispersion curve are clearly separated (Fig. 1.16).

Thus, neutrons give a description of the modes of vibration of crystals, even complex ones, and from that it is possible to obtain information about the interatomic forces. The main conclusion of this section, however, is that the global parameter, the heat capacity, does not in fact depend very much on the details of the dynamics of the vibrating atoms. The ancient Dulong and Petit's law of $3k_B$ per atom gives a result that is frequently only approximate but the discrepancy, if it exists, rarely exceeds 50%. Moreover, the sense of the discrepancy is known: the experimental value is almost always less than $3k_B$.

The rest of this book will show that there are few properties of solids that can be quantitatively predicted to this degree of approximation in such a simple manner.

Infra-red absorption in ionic crystals

We now return to the diatomic linear chain (Fig. 1.13) but replace the neutral atoms with ions of opposite sign, thus obtaining a one-dimensional model of an **ionic crystal** such as NaCl. All that has been said about the vibrations of the atoms in the chain remains valid for the ions.

If a beam of electromagnetic radiation is directed at this chain of ions, the vibrating electric field exerts opposite forces on adjacent positive and negative ions, which then tend to vibrate with opposite phases. The mode excited by the electromagnetic wave is the **optical mode** with zero wave vector (Box 6), because all the ions of the same sign vibrate in phase.

However, we know that a forced vibration only has an appreciable amplitude if there is resonance, i.e. if the frequency of the electromagnetic wave is equal to that of the optical mode. There is then strong absorption of the electromagnetic radiation, located in the far infra-red ($\lambda \approx 50$ μm), since the frequency is of the order of 10^{14} Hz.

Using quantum language, we can say an infra-red photon can be absorbed by the elastic wave if the excess energy which is given to it is exactly equal to its quantum energy, that of the phonon $\hbar\omega$.

Isolated molecules in solution or in the vapour phase are known to have infra-red absorption lines corresponding to the excitation of intramolecular vibrations. An analogous phenomenon explains the infra-red absorption in

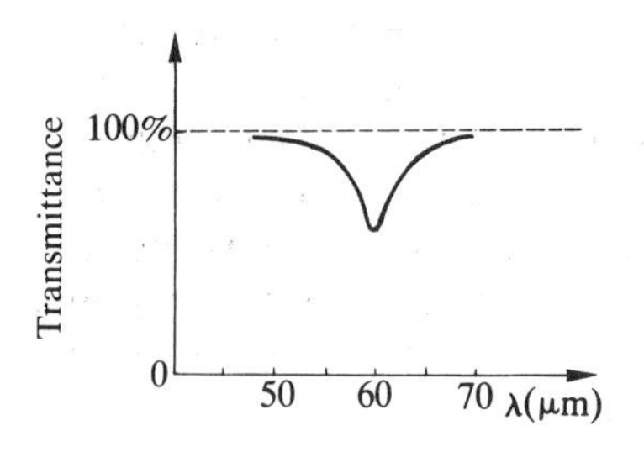

Fig. 1.17. Transmittance of infra-red radiation through a thin film (0.17 μm) of sodium chloride.

ionic crystals: these can be considered as giant molecules whose optical modes are the internal vibrations (Fig. 1.17).

This model enables several special features of the properties of ionic crystals in the infra-red to be explained. Thus, as shown in Box 6, the frequency ω_0 becomes smaller as the ionic mass increases. Now, it is in fact observed that the infra-red absorption band has a longer wavelength when the ions are more massive: it changes from 6 to 31 μm in going from LiF to KBr and reaches 150 μm with thallium iodide.

The excitation of the elastic wave in the crystal depends on the direction of the vibrating electric field and thus on the polarization of the infra-red wave. This is indeed what is observed experimentally.

When an electromagnetic wave encounters an absorbent medium, it is also reflected by the surface, an effect easily observed with metals. Ionic crystals have maximum reflecting powers in their infra-red absorption band. If a beam is reflected successively by several identical crystals, the intensities of the wavelengths centred on the absorption band are greatly increased in comparison with the rest of the spectrum. The reflected beam consists of what are sometimes called 'residual rays' (German *Reststrahlen*) which form the basis of a method used to isolate an infra-red band of long wavelength.

We have mentioned that, in the whole spectrum of elastic waves, those of very low frequencies can be excited by external sources. These are acoustic or ultrasonic waves whose technical possibilities limit the frequency to about 10^9 Hz. We now see that, at least for ionic crystals, there is another method of exciting elastic waves using electromagnetic radiation. However, this is only possible for waves with high frequencies around 10^{14} Hz. Between the two methods, there remains a range of frequencies extending from 10^{10} to 10^{13} Hz which exist spontaneously in the crystals due to thermal agitation. Their energy increases with increasing temperature and it is distributed between all the waves according to statistical laws (equipartition at high temperatures). In the intermediate range, however, *only spontaneous thermal waves exist*: there is no known method of exciting a particular wave as can be done in the ultrasonic and infra-red regions.

Thermal expansion

Solids increase in volume as their temperature is raised. The effect is very small, the relative expansion for a rise of 100 K being only about one part in a thousand for any solid. If the unit cell parameters are measured as a function of temperature, it is found that they increase by the same relative proportions as the macroscopic dimensions of the solid. The expansion is therefore due to a uniform displacement of all the atoms.

We use a simplified model to explain this phenomenon. Instead of considering a crystal, in which the atoms are bound to several neighbours, we shall base our argument on a pair of isolated atoms. These atoms are bound by a force which is the derivative of a potential. The minimum in this potential corresponds to the equilibrium distance a_0 between the pair of atoms. The atoms are in fact vibrating as a result of thermal agitation, so that their separation fluctuates around the value a_0.

We assume first of all that the interatomic potential is represented close to a_0 by a symmetrical curve, more precisely by a parabola (Fig. 1.18). The restoring force is then proportional to the amount by which the separation differs from a_0: the motion is said to be **harmonic**. For any amplitude of oscillation, and therefore for any temperature, the mean distance between the atoms remains constant at a_0, and there is therefore no expansion of the pair. It would be the same in a crystal if the vibrations of all the atoms were harmonic.

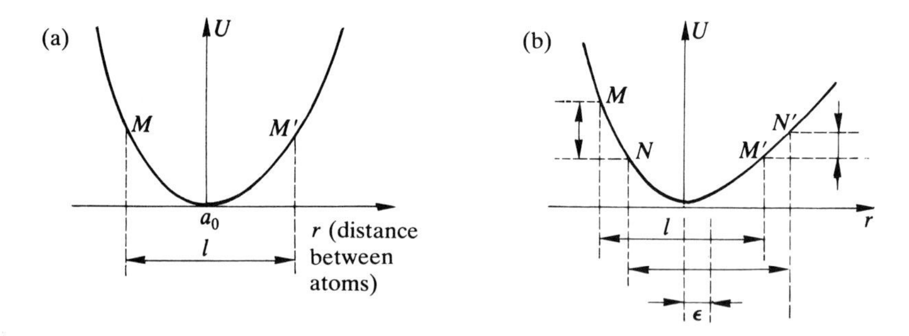

Fig. 1.18. Vibration of a pair of atoms around their equilibrium distance apart a_0 at 0 K: (a) with a symmetrical potential leading to harmonic vibrations; (b) with an asymmetrical potential leading to anharmonic vibrations.

However, this lack of expansion is contrary to experience and we must therefore abandon the basic assumption of the model, i.e. the parabolic shape of the potential curve near its minimum. This curve is, after all, clearly asymmetrical when taken as a whole (see Fig. 4.6, p.184): pushing the atoms closer together than the equilibrium separation a_0 requires the expenditure of more energy than pulling them further apart and this is still the case in the immediate neighbourhood of a_0.

Suppose the pair vibrates between M and M′ with an amplitude l centred about a_0. If the pair vibrates with the same amplitude but between N and N′ centred about $a_0 + \epsilon$, the system loses more energy from M to N than it gains from M′ to N′. The increase in the mean distance between the atoms reduces the vibrational energy of the pair: the equilibrium distance will correspond to the maximum possible reduction (a separation greater than this would raise the energy again).

If the temperature is raised, the displacement of the mean position from a_0 increases; it can be shown that this displacement is in fact proportional to the vibrational energy. The pair of atoms is thus 'expanded' by a rise in temperature. This expansion is a direct result of the **anharmonic nature** of the vibrations, i.e. of the asymmetry between forces brought into play as the atoms move closer or further apart. It is the same in a crystal, where the interatomic potentials are also anharmonic.

Using this as a basis for the theoretical calculation of the coefficient of expansion, the following result is obtained (Gruneisen's law): the coefficient is proportional to a parameter depending on the degree of the anharmonicity in the crystal and to the molar heat capacity, and it is inversely proportional to the atomic volume and the compressibility (i.e. the ratio of the relative decrease in volume to the increase in pressure producing it). At ordinary temperatures, the first three terms do not vary much from one metal to another and this means that it should be the compressibility that has the greatest effect. In fact, Table 1.4 verifies that soft metals have large coefficients of expansion while hard metals have small ones. Towards low temperatures, moreover, the coefficient of expansion decreases in a similar manner to the heat capacity and tends to zero at the absolute zero just as the heat capacity does.

Table 1.4 Coefficients of expansion for some solid elements

Substance	Coefficient of expansion ($10^{-6}K^{-1}$)	Compressibility ($10^{-12}Pa^{-1}$)
Cs	97	500
K	83	310
Li	45	86
Ag	20	9.9
Ni	13	5.4
W	4.5	3.1
C (diamond)	1.2	1.8

Some applications of expansion

As a general rule, all bodies increase in volume as their temperature rises. However, superimposed on this normally quite regular and universal expansion there are variations, either continuous or abrupt, which are manifestations of changes in the structure of the solid with temperature. That is why **dilatometry** is a simple and highly sensitive technique for

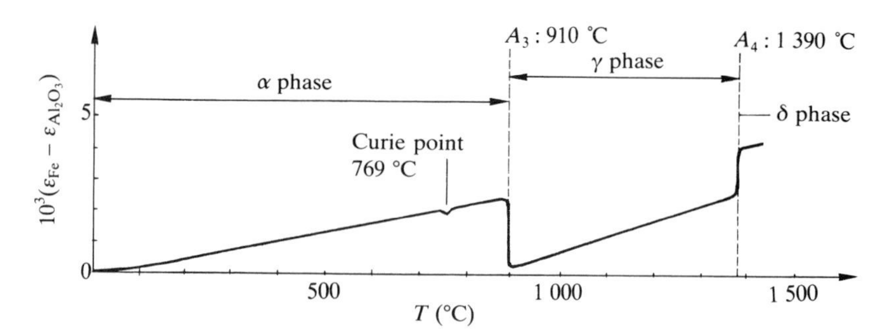

Fig. 1.19. Difference between the relative expansions of iron and aluminium plotted against temperature, showing evidence of the $\alpha \rightleftarrows \gamma$ and $\gamma \rightleftarrows \delta$ transformations in iron. The Curie point does not correspond to any change of lattice but is only revealed by a small dip in the curve.

detecting changes in the structure of metals and alloys as their temperature varies (Fig. 1.19).

For example, a nickel steel known as invar has a normal expansion which is exactly compensated by a structural contraction: invar is used in metrological apparatus where no change in dimensions must occur even though the ambient temperature may vary.

Thermal expansion is always small, but it plays an important role in a large number of techniques and must be taken into account in the design of any equipment. If two different solids are connected together in a single component, their coefficients of expansion must be very nearly equal for the component to withstand large variations of temperature. On the other hand, differences in expansion can be put to good use: a bimetallic strip, for example, can form a simple and robust temperature sensor.

Thermal conductivity

Until now, we have been dealing with solids in equilibrium at a temperature which is uniform throughout the body of the material. If at any given instant the temperature in a solid isolated from its external surroundings is not uniform, it equalizes spontaneously: heat is transported between the hotter and cooler parts without any transport of matter. This is the phenomenon we shall now investigate.

Consider a rod of cross-section S and length L, whose ends are put into contact with two thermostatic baths maintained at temperatures T_1 and T_2 (Fig. 1.20). A heat flux Q is permanently supplied to the hotter end and this is transported through the rod to the cooler thermostatic bath, which absorbs it. The whole system is in a steady state: quantitatively, it is

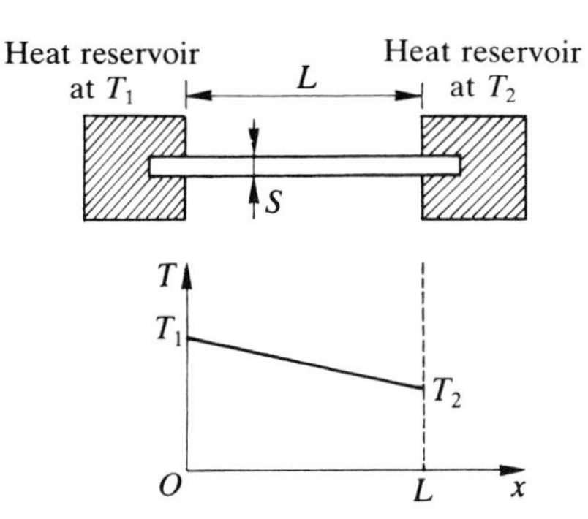

Fig. 1.20. Definition of the coefficient of thermal conductivity.

observed that the flow of heat per unit area, Q/S, is proportional to the temperature gradient $(T_1 - T_2)/L$ along the rod. The coefficient of proportionality is called the **thermal conductivity** K of the material:

$$Q/S = K(T_1 - T_2)/L.$$

In a rough and ready way, materials can be classified into two groups: good conductors, which are metals, and non-metallic materials (or electrical insulators) which are poor thermal conductors. Some materials have such a poor thermal conductivity that they are called thermal insulators. To give an idea of the orders of magnitude involved: at room temperature, the thermal conductivity of silver is $430\,\mathrm{W\,m^{-1}\,K^{-1}}$ and that of vitreous silica is $1.4\,\mathrm{W\,m^{-1}\,K^{-1}}$. Separating materials into two such well-defined classes is, of course, an oversimplification.

In metals, it is the free electrons which are responsible for the high thermal conductivity, just as they are for the high electrical conductivity. We shall leave this question for now and return to it after dealing with electrical properties (p. 76). We therefore confine our treatment for the moment to the thermal conductivity of the poor conductors that are electrically insulating crystals, i.e. ionic, covalent, or molecular crystals.

Conduction of heat by solids is an everyday phenomenon and the measurement of the coefficient of thermal conductivity can be carried out directly and easily. Its theoretical basis, however, is very complex and we should point out straight away that we do not know how to account for it quantitatively using calculations from first principles.

We analysed the thermal agitation of the atoms in a solid at a uniform temperature in terms of modes of vibration which are propagated throughout the whole crystal in all directions, the energy being equally distributed between all the various modes. In such a model, each wave propagates freely without interaction with any of the others. The amplitudes of the atomic oscillations are thus the same at all points in the crystal and the temperatures are therefore equal. It is impossible to explain how a temperature difference could be maintained inside such a crystal: in other words, the thermal conductivity would be infinite. We must therefore look for a way of introducing into the model a cause of the observed thermal resistance.

The problem is tackled by using two different approaches: in the first, the elastic waves used previously are preserved; in the second, the model adopted involves particles, the phonons we have encountered previously. Wave–particle duality is a fundamental concept in physics. The two models are both independent partial representations of a reality whose complete description is unknown to us. It is difficult to reconcile them: depending on the phenomenon to be explained, one will be more convenient or more fruitful than the other.

Thermal conductivity in terms of elastic waves

Because elastic waves cannot account for a finite conductivity while they remain independent, the idea of coupling between them is introduced. The independence of the modes of vibration follows from the assumption of linear elasticity in the crystal (i.e. the restoring force on the atoms being strictly proportional to their displacement). However, as we have already seen (p. 44), this is not exactly true: all crystals must be anharmonic because they expand with rise in temperature.

Consider a crystal through which a longitudinal elastic wave of wave vector $\boldsymbol{k}$ is travelling. Such a wave produces a succession of regions where the crystal alternately suffers expansion and contraction. Now consider a second wave of wave vector $\boldsymbol{k}'$. Because of the non-linearity of the crystal, the response of the atoms to the stress from the $\boldsymbol{k}'$ wave will be different in the two types of region. The $\boldsymbol{k}$ wave acts like a lattice producing a diffracted wave of wave vector $\boldsymbol{k}''$. Interference conditions lead to the relationship:

$$\boldsymbol{k} + \boldsymbol{k}' = \boldsymbol{k}''.$$

A certain redistribution of energy between the different modes thus occurs because of the coupling. However, this process, known as the normal process, does not help us in solving our problem: since the relationship between the wave vectors is analogous to the condition for the conservation of momentum, there is no resistance to the flow of heat.

There is, however, another possible process, known as an *Umklapp* process (umklapp = German 'flipping over'). Because of the periodicity in the atomic structure of the crystal, we know that wave vectors $\boldsymbol{k}$ and $\boldsymbol{k} + \boldsymbol{r}^*$, where $\boldsymbol{r}^*$ is any reciprocal lattice vector, are indistinguishable (p. 18).

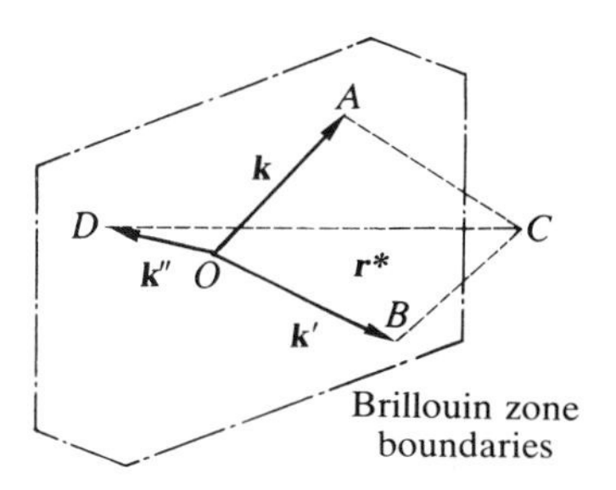

Fig. 1.21. Coupling between three modes of vibration by the Umklapp process: $\boldsymbol{k} + \boldsymbol{k}' = \overrightarrow{OC} = \boldsymbol{k}'' + \boldsymbol{r}^*$.

The interference condition can therefore also be written in the form:

$$\boldsymbol{k} + \boldsymbol{k}' = \boldsymbol{k}'' + \boldsymbol{r}^*.$$

The three vectors $\boldsymbol{k}$, $\boldsymbol{k}'$, and $\boldsymbol{k}''$ can be brought inside the Brillouin zone (Fig. 1.21): the wave $\boldsymbol{k}''$ and the waves $\boldsymbol{k}$ and $\boldsymbol{k}'$ are propagated in opposite directions, a point that is explained in Fig. 1.22 for the case of a linear chain. This type of coupling between waves, where there is no longer an analogy with the condition for the conservation of momentum, is capable of producing resistance to the flow of energy transported by the elastic waves.

There is also another source of resistance: diffraction of the waves can be produced by crystal imperfections, which alter the periodicity of the lattice: these may be local distortions, impurity atoms, or the presence of isotopes with their different atomic masses.

The phonon gas model

Corresponding to an elastic wave of frequency ω and wave vector $\boldsymbol{k}$ there is a phonon whose energy is the quantum energy of the wave $\hbar\omega$ and whose momentum is $\hbar\boldsymbol{k}$.[†] Corresponding to the whole set of vibrational waves in a crystal there is a phonon 'gas' in the volume bounded by the crystal surface, just as the black body radiation in a cavity is represented by a photon gas permeating the cavity. As with photons, the total number of phonons is indeterminate: phonons can disappear and others, in different numbers, can be created. (There is an important difference here with particles such as electrons or neutrons whose numbers are conserved.)

Phonons collide with each other, and this corresponds to the interactions between elastic waves. Between two successive collisions, the phonon travels a mean distance known as the mean free path Λ. The phonon has

[†]The wavelength λ of the wave associated with a particle of momentum $\boldsymbol{p} = m\boldsymbol{v}$ is given by the de Broglie relationship $\lambda = h/p$. Thus $p = h/\lambda = (h/2\pi)(2\pi/\lambda) = \hbar k$.

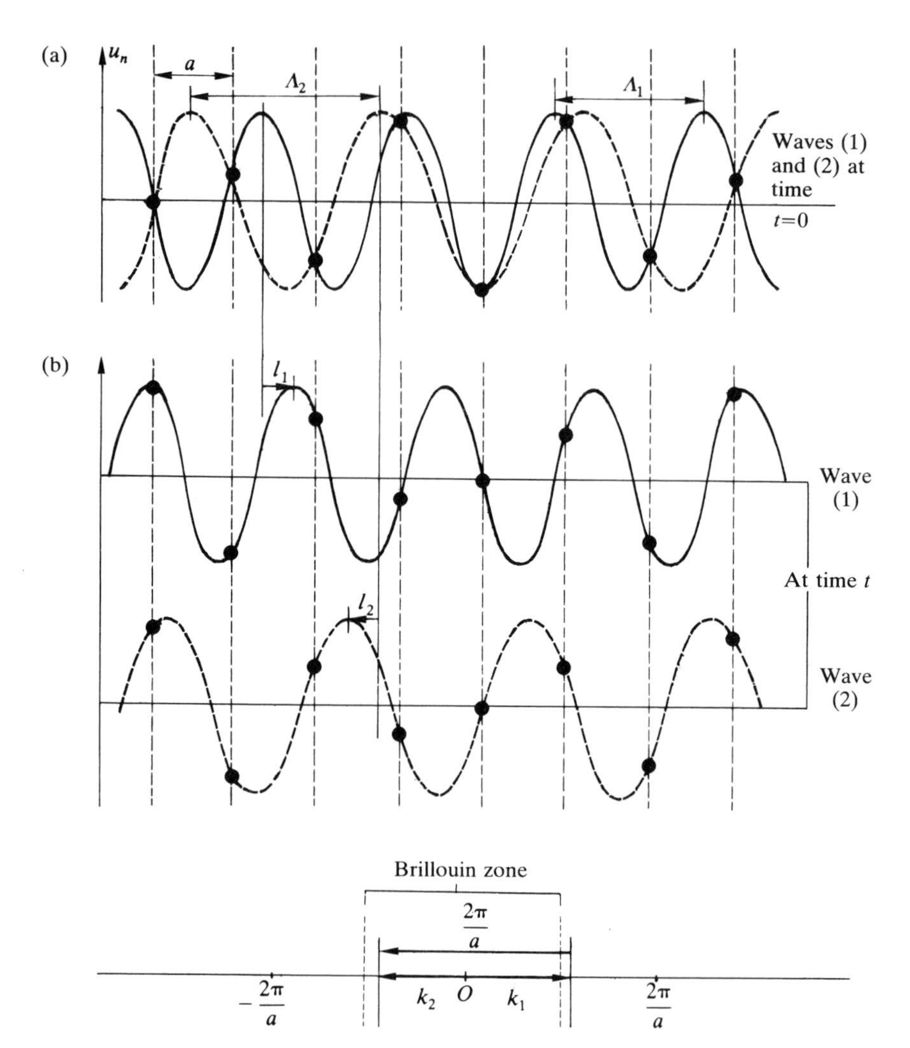

Fig. 1.22. The Umklapp process. (a) At time $t = 0$, the displacements are represented both by wave (1): $\Lambda_1 = 1.80a$, $k_1 = 2\pi/1.80a$ and by wave (2): $k_2 = k_1 - 2\pi/a = -2\pi/2.25a$, $\Lambda_2 = 2.25a$. (b) At time t, the displacements are represented by wave (1) having moved a distance l_1 to the right and by wave (2) having moved a distance l_2 to the left.

one feature which distinguishes it from a material particle: momentum is not always conserved in a collision since the term $\hbar \boldsymbol{r}^*$ can be involved in the conservation equation, $\boldsymbol{r}^*$ being *any* reciprocal lattice vector of the crystal.

Figure 1.23 shows how, in the case of collinear phonons, it is possible for two phonons $\boldsymbol{k}$ and $\boldsymbol{k}'$ to combine and give a phonon $\boldsymbol{k}''$ satisfying both the equation $\omega + \omega' = \omega''$ and $\boldsymbol{k} + \boldsymbol{k} = \boldsymbol{k}''$. In the case illustrated, momentum is

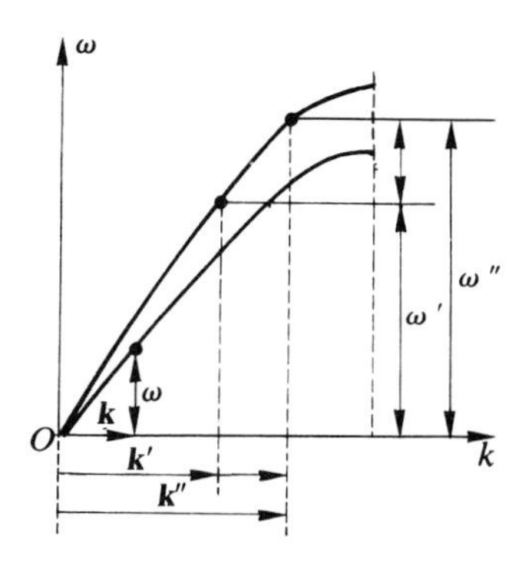

Fig. 1.23. Combination of three phonons such that $\boldsymbol{k} + \boldsymbol{k}' = \boldsymbol{k}''$ and $\omega + \omega' = \omega''$.

conserved (no resistance to the flow of heat; this is no longer true if $\boldsymbol{r}^*$ is different from zero). It will be noticed that the existence of a solution is related to the fact that the dispersion curve has several branches with a curvature of suitable sign and that transverse and longitudinal phonons must be involved.

In a gas, heat is transferred by the molecular motion: if, at the end of their mean free path, they are in a cooler region, they give their excess energy to the surroundings through collision with their neighbours. Using this picture, kinetic theory gives the following expression for the conductivity of a gas:

$$K = \frac{1}{3}\left(\frac{C}{A}\rho\right)\Lambda\upsilon$$

where C is the molar heat capacity; A is the molar mass; ρ is the density; υ is the mean molecular velocity; Λ is the mean free path of the molecules. This formula is established in Box 7.

Box 7

Calculation of the thermal conductivity of a gas

The thermal conductivity of a monatomic gas can be calculated using kinetic theory (SM, p. 51). The following treatment gives an approximate derivation.

Assume that, in the vessel containing the gas, the temperature is not uniform but varies along a single spatial direction x. All the points lying in a plane with coordinate x are then at the same temperature $T(x)$ but the temperature varies from one value of x to another. At the coordinate x, there is a local temperature gradient $\mathrm{d}T/\mathrm{d}x$. We assume that, just after a collision, a gas molecule possesses kinetic energy corresponding to the temperature at the point where it is situated:

$$E = \tfrac{1}{2}mv^2 = 3k_BT/2.$$

It transports this energy from one point to another in the vessel as long as it does not suffer another collision. Energy transfer in the gas thus only occurs during the course of collisions between molecules.

Corresponding to these microscopic transfers of energy there is a macroscopic flux of energy or heat Q. This heat flux, defined algebraically across a plane at coordinate x, is calculated by adding together the energies of the molecules which cross unit area in this plane per unit time in the positive x direction and by subtracting the energies of the same number of molecules which cross it in the negative direction:

$$Q = \nu(\mathrm{E}_+ - \mathrm{E}_-)$$

where ν is the number of molecules crossing unit area of a plane per unit time in a given direction.

If all the molecules moved only along the x axis, ν would be equal to the number of particles contained in a slice of unit cross-section and thickness v, since in unit time all these molecules would have travelled a distance v and thus crossed the plane (Fig. 1.24). We should therefore have that

$$\nu = nv$$

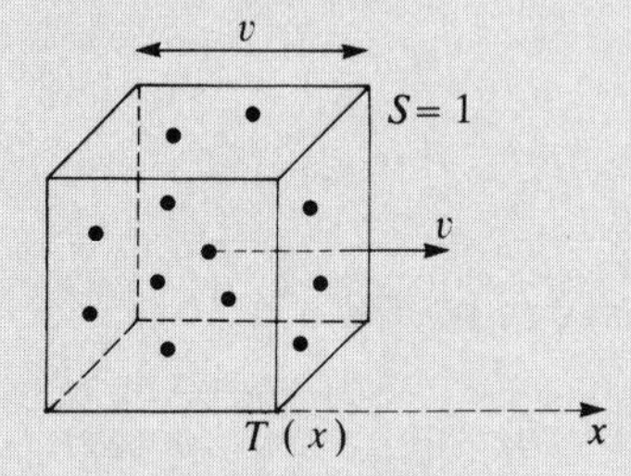

Fig. 1.24. Illustration of the method of calculating the thermal conductivity of a gas.

where n is the number of particles per unit volume and v their mean speed. A more rigorous calculation, which takes into account all the possible directions of the velocities with respect to the plane, gives:

$$\nu = nv/4.$$

It remains to evaluate the energies E_+ and E_- transported by the particles crossing the plane in the direction of increasing and decreasing x respectively. E_+ corresponds to the temperature the molecules had at the time of their last collision. If, between two collisions, the molecules travel a distance Λ (the mean free path), E_+ corresponds to the temperature at a coordinate $x - a$, less than x, where a is of the same order as Λ. A calculation including all possible directions of the velocity shows that $a = 2\Lambda/3$. We therefore have:

$$E_+ = \frac{3}{2} k_B T \left(x - \frac{2}{3} \Lambda \right).$$

In the same way:

$$E_- = \frac{3}{2} k_B T \left(x + \frac{2}{3} \Lambda \right).$$

From these, we deduce that:

$$Q = \frac{3}{8} n \upsilon k_B \left[T\left(x - \frac{2}{3}\Lambda \right) - T\left(x + \frac{2}{3}\Lambda \right) \right].$$

If we are interested in macroscopic variations in the coordinate x much greater than Λ, the temperature difference between $x + 2\Lambda/3$ and $x - 2\Lambda/3$ can be expressed in terms of the mean local temperature gradient with respect to x, i.e. $\mathrm{d}T/\mathrm{d}x$. Hence:

$$Q = -\tfrac{1}{2} n \upsilon k_B \Lambda \mathrm{d}T/\mathrm{d}x.$$

The minus sign indicates that the heat flow is in the opposite direction to the local temperature gradient and thus tends to make the temperature more uniform. The coefficient of $\mathrm{d}T/\mathrm{d}x$ corresponds by definition to the thermal conductivity K, so that

$$K = \tfrac{1}{2} n \upsilon k_B \Lambda.$$

The number of atoms per unit volume n and the density ρ are related by

$$\rho = nA/N_A$$

where A is the molar mass and N_A is Avogadro's number. The value of K can then be expressed as

$$K = \tfrac{1}{2} \rho \upsilon (N_A k_B / A) \Lambda.$$

Finally, we remember that the molar heat capacity C is

$$C = 3R/2 = 3N_A k_B/2$$

so that

$$K = \frac{1}{3}\left(\frac{C}{A}\rho\right)\Lambda v.$$

If, allowing ourselves some latitude, we use this formula for the case of a phonon gas, with the velocity v of the phonons becoming the speed of sound in the material, the mean free path can be derived from experimental data and we end up with results that are not too absurd. Thus, for NaCl at 0°C, $\Lambda = 2.3\,\text{nm}$ and for quartz at 0°C, $\Lambda = 4\,\text{nm}$.

This justifies the use of the formula for a quick calculation of orders of magnitude and for qualitative arguments. If a more refined analysis is required, the calculations with both wave and phonon models are very complicated and, as has already been mentioned, do not enable thermal conductivities to be evaluated from first principles without the introduction of empirical parameters.

Qualitative explanation of some experimental results

Variation of thermal conductivity with temperature

At high temperatures ($T > \Theta_D$), the heat capacity is almost constant. The coefficient K is therefore proportional to the mean free path, Λ. This is inversely proportional to the number of phonons present, and since the number of phonons excited as the temperature rises is proportional to T, it follows that Λ, and thus the conductivity, is proportional to $1/T$.

Near the Debye temperature, the heat capacity begins to decrease. The conductivity is therefore less than the value extrapolated from high temperatures by using the $1/T$ law.

At very low temperatures, few phonons are excited, but the mean free path does not increase indefinitely, firstly because it is limited by imperfections in the lattice and, secondly, even if the lattice were perfect, it is limited by the size of the specimen. (This is reminiscent of the 'molecular regime' or 'Knudsen regime' in rarefied gases; SM, p. 48.) It has been observed that the conductivity at low temperatures, measured on cylindrical rods, depends on the diameter of the rod. If it is assumed that Λ becomes constant at low temperatures, the conductivity is proportional to the heat capacity: it is zero at the absolute zero and varies with T^3 at low temperatures.

In this way, we can account for the shape of the curve (Fig. 1.25) of the conductivity due to phonons (or, in other words, due to the lattice, so as to distinguish it from the conductivity of metals due to electrons).

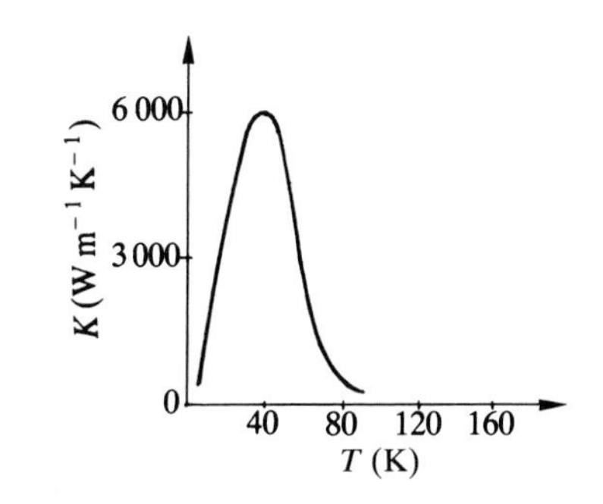

Fig. 1.25. Thermal conductivity of sapphire (crystalline alumina or corundum) and its variation with temperature.

Comparison of the conductivities of various solids

It is remarkable that the thermal conductivity of a crystal such as alumina (sapphire) at the temperature of its maximum (40 K) exceeds that of silver at room temperature: 6000 W m^{-1}K^{-1} compared with 420 W m^{-1}K^{-1}, which means that the mean free path of the phonons must be long. This can be explained by the perfection of the crystal and by the fact that, because of the high Debye temperature (very strong interatomic bonds), the high frequency phonons necessary for the Umklapp process are not excited in very large numbers. Diamond also has good thermal conductivity at low temperatures, which accounts for an ancient custom of jewellers: to distinguish a sapphire or ruby from glass, they place it on their tongue: the gemstone gives an impression of being cold, unlike the glass.

On the other hand, materials in which the crystallization is imperfect such as polymers are very poor conductors of heat. The mean free paths of the phonons here are very short.

Thermal insulators used for practical applications are of course produced from very poorly conducting solids, but their insulating properties arise largely from their porous structure. Thin solid layers surround cavities filled with air, which is a very good insulator as long as there is no convection. (This is why woollen fabrics are very good insulators.)

2

Electrical properties of solids

What happens when a solid is placed in an electric field? In answering this question, we shall not be concerned with experimental details nor with the particular properties of various solids and the many applications that depend on them: these aspects are matters for texts on electricity or electronics. The aim in this chapter is rather to account for the electrical behaviour of solids in terms of their atomic structure. We shall see that we are led to refinements of the models used in more elementary accounts of the structure of matter (e.g. SM, Chapters 1 and 4).

Observation shows that solids can be broadly classified into two groups, electrical **conductors** and **insulators**, depending on whether or not an electric field causes a current to flow in them. The property of solids giving the clearest indication of their ability to conduct electricity is their resistivity, a quantity showing enormous variations from one substance to another: the ratio of the resistivities of quartz and copper is 10^{20} (which is of the same order of magnitude as the ratio of the sun's diameter to that of an atom). The resistivities of real materials are in fact spread over the whole of this range, with no large intervals that are completely empty. In spite of that, it is possible to make a clear separation of substances into good conductors and good insulators, with the exception of some intermediate solids which we return to later because of their importance.

A simple relationship between the electrical conductivity and the type of atomic bond is immediately apparent. Good conductors are metals, with crystal structures in which some of the electrons are free, i.e. not strongly bound to one or more atoms. Molecular, ionic and covalent crystals, on the other hand, are insulators.

This is an elementary yet fundamental statement and it bears closer examination. Consider an insulating solid such as an ionic crystal (SM, p. 78), in which all the electrons are concentrated in either positive or negative ions. An applied electric field exerts a force on every one of the electrons but, even if the field is extremely strong, the force is still very weak compared with that binding the electron to the positive nucleus. Although its motion may be slightly modified, the electron remains bound to its nucleus, so that no electron current can flow in the crystal. Furthermore, there is no way that the ions as a whole can move since there is nowhere for them to go in the close-packed structure of the crystal. The same is true for molecular and covalent crystals.

High conductivity can only occur when electrons are almost free, i.e. when they are so weakly bound to the atomic nuclei that they can be set in motion by the external field. This provides the very simple basis for the theory of metallic conduction. However, simple classical models are inadequate when it comes to providing a *quantitative* explanation of the properties of metals, and this is even more so for semiconductors, whose conductivities are intermediate between those of conductors and insulators: here it is absolutely essential to introduce the concepts of quantum physics.

Although an electric field applied to a solid insulator does not produce a current, it does have other effects, and these give rise to important properties both from the fundamental point of view and from that of technical applications. When these are being discussed, we prefer the term **dielectric** to 'insulator', which has a purely negative connotation.

Dielectrics

The application of an electric field to a dielectric produces effects of several different types. In discussing these, we shall assume that the dielectric is a perfect insulator or, more accurately, we shall neglect the possible existence of a very small residual conductivity.

1. When the previously empty space between the plates of a capacitor is filled with a dielectric, it is well known that the capacitance increases. Let us assume for the moment that the dielectric is electrically isotropic, which is true for crystals belonging to the cubic system, for polycrystalline dielectrics and for amorphous materials. The capacitance is then increased by a factor ϵ_r, the dielectric constant or **relative permittivity**. This is a pure number, always greater than 1 and normally with a value in single figures: only for a few exceptional solids is ϵ_r greater than 10.

2. Unlike metals, dielectrics may be **transparent** to light waves and, more generally, to electromagnetic waves, at least over a certain range of wavelengths. This is revealed by their colour: it is a common observation that insulators are often coloured and never have a metallic lustre. The action of the electric field of the wave on the dielectric is revealed by the propagation of the wave: its velocity of propagation, c/n, is less than its velocity in a vacuum, c. The **refractive index**, n, is related to the relative permittivity through the Maxwell expression $n^2 = \epsilon_r$.

3. When subjected to a high frequency electric field, the dielectric in a capacitor absorbs energy to an extent that varies from one substance to

another: this is revealed by the dissipation of heat in the capacitor. Another aspect of this is that an electromagnetic wave passing through the dielectric is not transmitted without modification: there is a partial absorption of energy from the wave during its propagation.

4. In crystals other than those belonging to the cubic system, the action of the electric field depends on its direction with respect to the crystal axes. Thus, the permittivity of a dielectric sheet between capacitor plates varies with the orientation of its surfaces in relation to its crystal axes. It is the same for the propagation of a wave through the sheet. The complex effects observed (birefringence, optical activity) and their diverse applications are the subject of an important branch of optics: crystal optics.

It can be seen from this brief survey that a dielectric which is a perfect insulator is very far from being passive under the action of an electric field. The phenomenon underlying all the facts we have just described is that of the polarization of the dielectric. This forms the link between the distribution of the electric charges in the material and its macroscopic properties.

Dielectric polarization

Dielectrics in zero electric field

The atoms in a crystal consist of charged particles, the positive nuclei and the negative electrons. Whatever their arrangement, the total charge in a unit cell of the crystal is zero.

Moreover, for the large majority of substances, the centroid of the positive charges in the molecule or unit cell coincides with that of the negative charges. The electric dipole moment of the molecule or unit cell is then zero and the molecule or unit cell is said to be **non-polar**. Such a

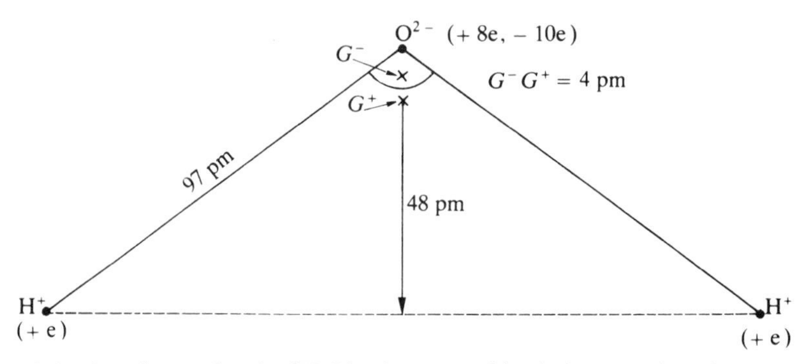

Fig. 2.1. A polar molecule (H_2O): the centroid of the negative charges is not located at the centre of the O^{2-} ion because the electron cloud is perturbed by the H^+ bonds. The dipole moment of the molecule is $10 \times 4 = 40$ electron-pm (1 picometre $= 10^{-12}$m).

molecule or crystal possesses no tendency to change its direction in an electric field.

The opposite, less common, case is that of a **polar** molecule, which does possess an electric moment in the absence of an applied electric field. For example, the CO_2 molecule is linear and has a centre of symmetry, so it is non-polar; the water molecule H_2O illustrated in Fig. 2.1, on the other hand, has an electric dipole moment. It is worth noting, however, that ice crystals are non-polar (SM, pp. 77–8) since the four molecules in the unit cell are so arranged that the total moment is zero.

Polar crystals, in which the centroids of the positive and negative charges do not coincide, are less common, but we shall nevertheless be discussing them later because of their important properties.

Dielectrics in an applied electric field

When a steady electric field is applied to a non-polar dielectric, the positive and negative charges in the unit cell are subjected to forces tending to move them in opposite directions. As we have already said, these forces are insufficient to detach an electron from its atom or ion, but they will produce a slight displacement of the negative charges in relation to the positive charges, a change which is opposed by the forces both within the atom or ion and between ions. When equilibrium is attained, the centroids of the charges of opposite sign ($G+$ and $G-$ in Fig. 2.2(b)) no longer coincide, so that in an electric field the unit cell of the crystal possesses an **induced electric dipole moment**. The distance $G+G-$ is proportional to the electric field E if the restoring forces are proportional to the displacements. Thus, to a good approximation, the dipole moment of the unit cell induced by the electric field is proportional to the field, as is the corresponding macroscopic property of the crystal: its **electric polarization**, the vector $\boldsymbol{P}$. This means that

$$\boldsymbol{P} = \epsilon_0 \chi \boldsymbol{E}$$

where χ is the **electric susceptibility**, a dimensionless number lying between

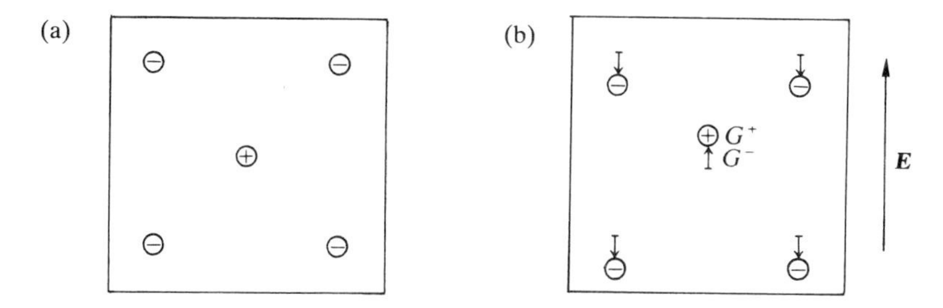

Fig. 2.2 Projection of the unit cell of a non-polar crystal: (a) no applied electric field; (b) in an electric field $\boldsymbol{E}$.

0 and 10 for most solids. The quantity ϵ_0 is the **permittivity of free space**, with a value that depends on the system of units being used. In SI units, $\epsilon_0 = 8.85 \times 10^{-12}\,\mathrm{F\,m^{-1}}$.

When the laws of electrostatics are applied to such a system, they account for the increase in the capacitance of a capacitor due to a dielectric and show that the relative permittivity ϵ_r is equal to $1 + \chi$.

Take NaCl as an example. It is obvious from the regular alternation of positive and negative ions that the dipole moment of the unit cell is zero. When an electric field is applied, the crystal becomes polarized and its susceptibility χ is 4.8. Thus, in an electric field of $10^3\,\mathrm{V\,m^{-1}}$, the polarization P is $42 \times 10^{-6}\,\mathrm{C\,m^{-2}}$ and the dipole moment of the unit cell with a volume V of $175 \times 10^{-30}\,\mathrm{m^3}$ is $m = 7 \times 10^{-33}\,\mathrm{C\,m}$. Since the unit cell contains $4Na^+$ ions and $4Cl^-$ ions, the charges are $+4e$ and $-4e$. The separation of the centroids of these charges is therefore $m/4e$ or 1.1×10^{-5} nm. This shows that the polarization hardly deforms the crystal at all.

If the crystal is anisotropic, the atomic restoring forces depend on the direction of the displacements. Consequently, the displacements of the atoms, and hence the polarization vector, may not have the same direction as the electric field $\boldsymbol{E}$ (Fig. 2.3), although the modulus of $\boldsymbol{P}$ is still proportional to the modulus of $\boldsymbol{E}$. The proportionality between $\boldsymbol{P}$ and $\boldsymbol{E}$ is no longer expressed as a scalar, but as a **tensor**: a tensor, moreover, that has the same symmetry properties as the crystal. All the optical properties specific to crystals, particularly their birefringence, follow from this fact.

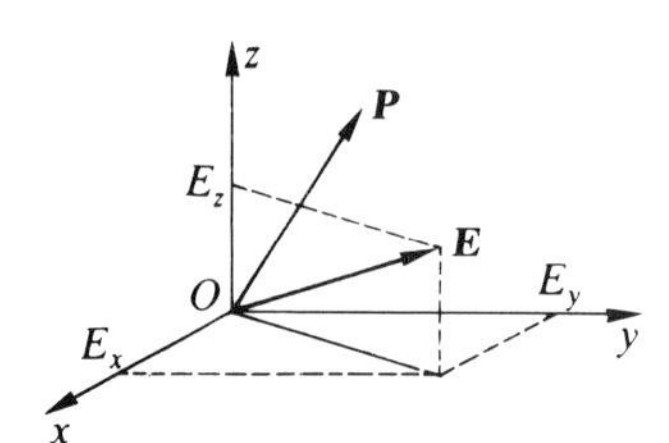

Fig. 2.3. Polarization of an anisotropic crystal. In any crystal, there exists an orthogonal system of axes such that:
$\boldsymbol{P} = a\boldsymbol{E}_x + b\boldsymbol{E}_y + c\boldsymbol{E}_z$.

The mechanisms of electric polarization

Electric polarization is the result of the relative displacements of charges. There are several possible ways in which this may occur, each giving rise to different dielectric properties: the alignment of polar molecules as a whole, the relative displacement of ions, or the deformation of atoms.

The alignment of **polar molecules** cannot occur in a crystal since the position of the molecules is fixed, but it can in a liquid, which has a disordered structure. In the absence of a field, the molecular dipole moments are tightly packed against each other without any order and with random directions which are nevertheless continuously varying because of the thermal agitation. The resultant dipole moment is therefore zero. When an electric field is applied, a couple is exerted on the molecules tending to align their dipole moments along the field direction. This tendency increases as the field increases and as the thermal agitation (and thus the temperature) decreases. As a consequence of this, there is a resultant dipole moment proportional to the field, although the molecular alignment always remains very far from being perfect.

Thus, for water in a field of $100\,\mathrm{V\,m^{-1}}$, the polarization is $10^{-5}\,\mathrm{C\,m^{-2}}$, whereas it would be $0.3\,\mathrm{C\,m^{-2}}$ if all the molecules had their dipole moments parallel to the field. Not only that, but water in its liquid state is an exceptional dielectric with a very high relative permittivity ($\epsilon_r = 80$), so that the alignment will be even less perfect in most substances.

There is, however, a special class of crystal in which the alignment of polar molecules is possible: *plastic crystals*. In these, the molecular centres have fixed positions but the molecules as a whole are free to take up one of several directions. In an electric field, a molecule tends to choose the orientation giving it the minimum energy.

The two important mechanisms of polarization in **non-polar crystals** are the relative displacement of complete ions (ionic polarization) and the deformation of ions or atoms (electronic polarization). These mechanisms can be distinguished from each other by subjecting the crystal to an alternating electric field. The technique involves the measurement of electric susceptibility as a function of the frequency of the applied field.

Electronic polarization

In this case, the atom or ion acquires a dipole moment by virtue of the relative displacement of the electrons and the nucleus. Since the inertia of electrons is very low because of their small mass, variations in electronic polarization can follow the vibrations of the field up to very high frequencies ($10^{16}\,\mathrm{Hz}$) corresponding to waves in the ultraviolet region.

Classical theory represents the complex motion of the electrons in the atom by a set of harmonic oscillators with natural frequencies ranging from the visible to the ultraviolet. When subjected to an alternating exciting force, the oscillators give rise to effects due to resonance: if the frequency of the applied field coincides with the natural frequency of an oscillator, the susceptibility increases and, at the same time, so does the energy absorbed from the wave by the atom. This is the origin of **anomalous dispersion**, a large variation in the refractive index over a narrow range of frequencies (Fig. 2.4).

Ionic polarization

Here, the ion is displaced as a whole and, since its mass is about 1000 times that of an electron, the natural frequencies of the ionic oscillations are

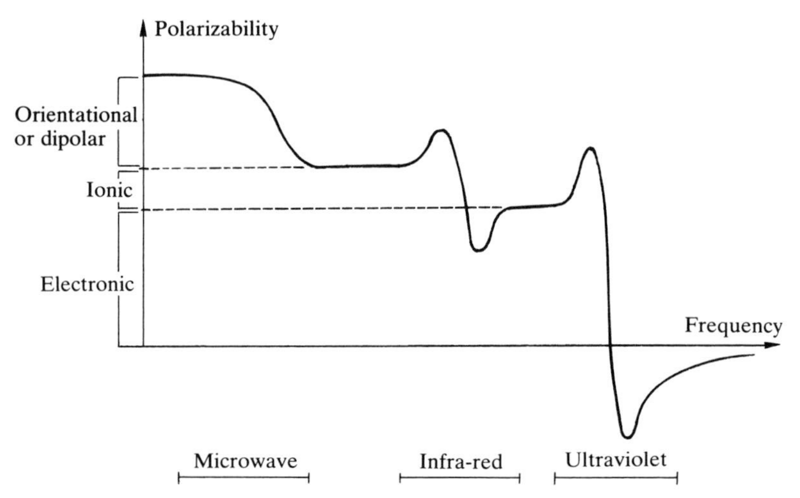

Fig. 2.4. Variation of the polarizability of a solid with frequency of the applied electric field.

much lower than those of electrons. In fact, the ions cannot follow high frequency variations and the ionic susceptibility then becomes negligible. This transition occurs in the visible region or in the infra-red. We have already indicated (p. 42) that, for alkali halide crystals, the optical modes for ionic vibrations are located in the far infra-red.

Piezoelectric crystals

These are crystals that become polarized when subjected to mechanical stress, whether compressive or tensile: **quartz** is the best known example. The importance of piezoelectricity lies in its many modern applications: quartz watches, ultrasonic generators, and so on.

Consider a crystal whose unit cell, when at rest and not subjected to an applied electric field, has a zero dipole moment. Its positive and negative charges are distributed in such a way that their centroids coincide. If pressure is applied to a thin slice cut from the crystal, it suffers a very small elastic strain (change in thickness of the order of 1/1000) proportional to the pressure. On an atomic scale, the unit cell suffers the same deformation and this produces small displacements of the atoms in it relative to each other.

If the unit cell has a centre of symmetry, the crystal remains centrosymmetric after the strain and therefore acquires no dipole moment (Fig. 2.5). It is quite different if the unit cell lacks a centre of symmetry: the displacement of the positive and negative ions may separate the centroids of the two types of charge. The crystal then becomes polarized

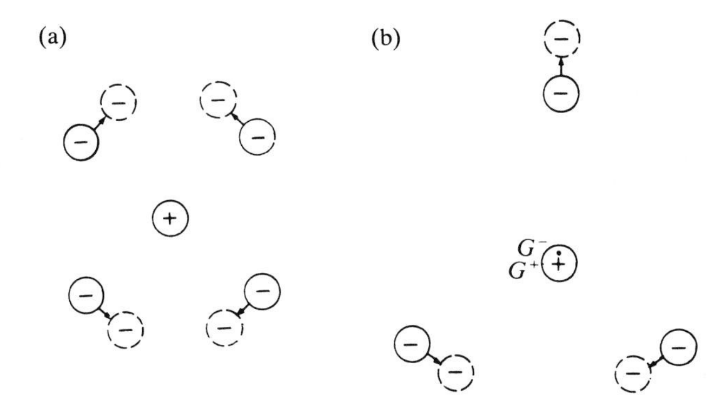

Fig. 2.5. The unit cell, when deformed by a mechanical stress, is (a) not polarized if it has a centre of symmetry, but (b) may be polarized if it lacks such a centre.

and, since the displacements are proportional to the applied pressure, so is the polarization. If the surfaces of the slice are covered with metal films (Fig. 2.6), a potential difference proportional to the applied pressure appears between them.

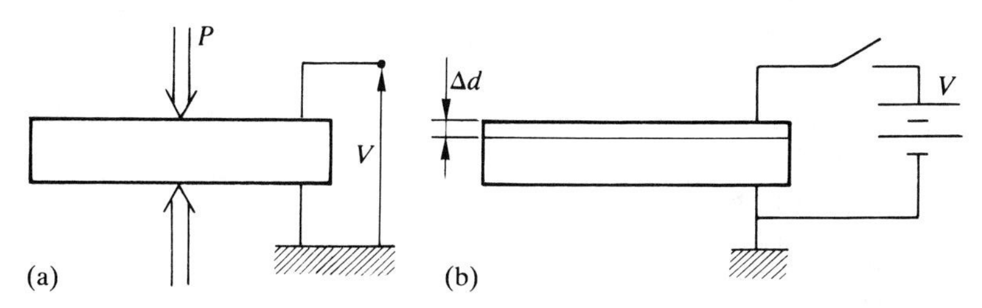

Fig. 2.6. (a) The direct piezoelectric effect: polarization charges created by pressure; (b) the inverse effect: the crystal slice is mechanically deformed when a voltage is applied across it.

Although the elastic strain in a crystal is always very small, there are fairly rare cases in which the crystal structure is such that the electric polarization induced is large enough to be detected and used. For instance, if the 10 cm by 2 cm surfaces of a suitably oriented thin quartz slice of thickness 0.5 mm are pulled apart by a force equal to the weight of 0.5 kg, opposite charges of 2×10^{-9} C appear on them.

There is also an inverse piezoelectric effect. If a voltage is applied across the faces of the crystal slice, it experiences an electric field and becomes polarized. The ions of opposite sign are displaced by electric forces acting in opposite directions, producing a deformation of the unit cell and thus a change in the shape of the slice. Its thickness varies by an amount proportional to the applied voltage.

If the applied voltage is alternating, the thickness of the slice oscillates at the same frequency, thus exciting elastic vibrations in it. However, the amplitude of these vibrations will not be large unless a natural mode is excited. The electrical and elastic vibrations are therefore coupled: the impedance of the electric circuit is controlled by the frequency of the natural mode of the slice. Now the vibrations in quartz have a very low decay constant and the resonance is therefore very sharp (high Q-factor). As a result, the quartz slice acts as a frequency stabilizer and it is this property that has led to the success of the quartz watch. In general, the natural frequency of a quartz slice varies with temperature. Fortunately, it is possible to cut a slice from the crystal in such a direction that the natural frequency is not sensitive to temperature variations around normal ambient temperatures. This means that a quartz oscillator can be stabilized to within 1 in 10^8. The electric circuit only functions in an extremely narrow frequency band centred on a value determined by the quartz crystal: a small rod 4 mm in length and 2 mm in diameter gives a watch an operational stability of about 1 second per year.

If a liquid is brought into contact with the vibrating surface of the quartz, elastic vibrations will be produced in it. The frequency of the waves transmitted through the liquid can be changed by altering the thickness of the slice. Quartz oscillators form practical ultrasonic sources in the frequency range from 10^7 to 10^{10} Hz. The availability of such sources has led to the development of many applications in the ultrasonic field. Underwater communication is one such field: sonar is an ultrasonic form of radar for detecting an obstacle that reflects ultrasonic waves and for measuring its distance from the source. We might also mention the examination of the human heart using echocardiography.

An experimental test to ascertain whether a small crystal is piezoelectric is quite easy to perform. Such a test, carried out on a macroscopic scale, thus enables a crystallographer to discover whether or not, on the atomic scale, the unit cell of the crystal has a centre of symmetry.

Polar crystals

A unit cell contains a total positive charge Q and a compensatory negative charge $-Q$. However, in a polar crystal, the atomic arrangement is such that the centroids of these two charge systems do not coincide even when there is equilibrium and no applied electric field. If the centroids are a distance l apart, the spontaneous polarization of the crystal is Ql/V, where V is the volume of the unit cell.

A crystal is only polar if its unit cell lacks a centre of symmetry. If it does have such a centre, the symmetry operation of inversion through it would by definition leave the unit cell unchanged, but the electric dipole moment would be reversed: the latter must therefore be zero.

It seems natural to assume that a polar crystal would be the source of an electric field around it, just as a permanent magnet is the source of a magnetic field which finds many applications. In practice, there is no such field: the electric polarization does not reveal itself in an obvious way, for the following two reasons.

1. In the first place, the energies involved in normal electrostatic phenomena are very small compared with those associated with magnetic effects. Consequently, whereas forces of magnetic origin are easily detectable in comparison with other forces such as weight, friction, etc, those arising from electrostatic effects are not.

It should be remembered that electric polarization is at best only a small effect because the ions are close-packed in the crystal and can only be displaced through extremely small distances.

2. The analogy between electric polarization and magnetization of materials is formally complete if 'magnetic charges' are invoked to correspond to the electric charges, but positive and negative electric charges have a separate existence, whereas the basic magnetic elements are *dipoles*.

This explains why an effect that occurs with polarized solids has no counterpart with magnets. The surfaces of a polarized slice attract free charges existing in the solid (impurities) or present in the surrounding atmosphere in the form of the ions. The electric dipole moment of these 'real' charges very quickly neutralizes the intrinsic polarization (Fig. 2.7), thus making it impossible to detect the effective polarization of the solid.

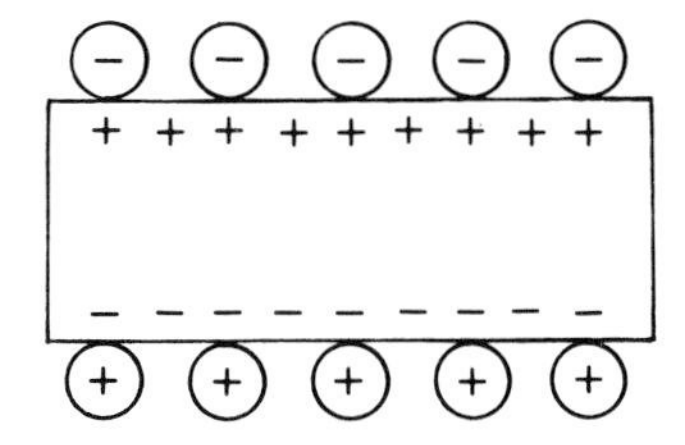

Fig. 2.7. Neutralization of polarization charges by free external charges.

Pyroelectric crystals

Nevertheless, the spontaneous polarization is revealed in certain crystals where it shows a large variation with temperature. Tourmaline, a natural mineral with the general formula (Na, Li, Ca) (Mn, Mg, Fe, . . .)$_9$ (OH, F) $(BO_3)_3$ (Si_6O_{18}), is an example. Its atomic structure changes with temperature in such a way that the spontaneous polarization decreases as the temperature falls: at −250 °C, it is no more than 7 per cent of its value at room temperature. When tourmaline is cooled, its polarization changes immediately, whereas the neutralization by external charges takes place only slowly. The polarization of the tourmaline is thus temporarily revealed, but the effect has not been used for any important applications.

Electrets

Electrets are made from certain polymers such as teflon (polytetrafluorethylene) in the form of thin sheets with thicknesses between 10 and 50 μm. They have the unusual property of keeping their polarization and/or their 'real' charges permanently. The sheet is polarized in a strong field and at a high temperature (above 150 °C), and then cooled in the field. The dipoles that are oriented at the higher temperature remain fixed at ordinary temperatures. The deposition of the charges on the sheet is achieved by an electric discharge in the air in contact with it or by electron bombardment. The permanence of the charges implies that the dielectric has an extremely high resistivity.

Electrets are mainly used in the manufacture of microphones. The charged polymer sheet, metallized over one surface, is stretched in front of an insulated metal plate from which it is separated by a thin layer of air. Sound waves falling on the elastic sheet deform it and the resultant displacements produce a voltage between the two electrodes whose variations faithfully follow those of the position of the sheet and therefore of the pressure. This is a simple type of microphone compared with others: it has a high sensitivity and has the great advantage of being able to function indefinitely without any external electrical supplies.

Ferroelectric crystals

These are unusual, and involve only a small number of crystal species, but their special properties find so many applications that we are justified in discussing them. However, we shall only deal with one fairly typical example, that of barium titanate ($BaTiO_3$).

Above 120 °C, barium titanate is a crystal with a very simple and highly symmetrical structure: the unit cell is cubic, with a titanium atom at the centre, a barium atom at the corners, and the three oxygen atoms at the centres of the cube faces (Fig. 2.8).

At 120 °C, a change of phase occurs: the cubic unit cell becomes tetragonal (a rectangular parallelepiped with a square base) having a height that is slightly greater than the side of the square base ($c/a = 1.04$). What is more significant, however, is that the atoms do not keep the same positions below 120 °C that they had in the cubic phase: the positive ions, Ba^{++} and Ti^{++}, are displaced along the c-axis relative to the oxygen ions.

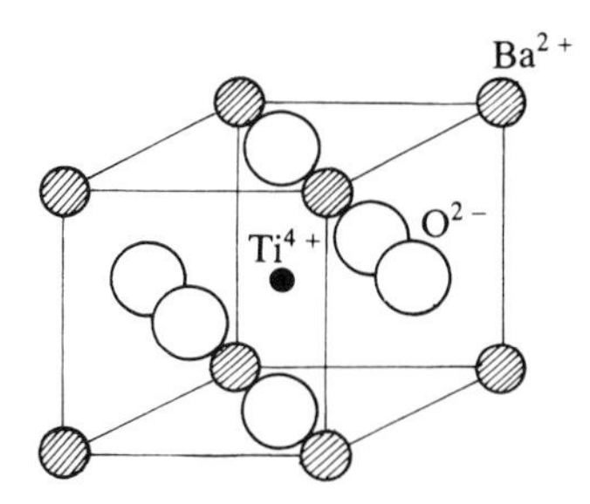

Fig. 2.8. Cubic unit cell of barium titanate.

The crystal thus becomes polarized with a dipole moment directed along *c*. The ionic displacements are very small, only about 1/1000 of the interatomic distances.

As a general rule, the structure with the highest symmetry is the stable one at high temperatures, with the ions vibrating about their mean positions. Below the transition temperature, the symmetry is 'broken' and it is the slightly distorted arrangement which becomes the more stable one. However, more than one orientation is possible for the new phase: for example, the titanium atom in the unit cell can equally well be displaced in six directions (along the three cubic axes in either direction).

The different unit cell structures are not, however, mixed up at random. The phase transformation is a cooperative phenomenon which creates domains of uniform structure of quite large dimensions (of the order of $10\,\mu m$ or even much more). There are several methods of obtaining a picture of these domains on the surface of a titanate crystal and remarkable

Fig. 2.9. Domains in barium titanate. Observation of a thin section under the polarizing microscope reveals domains with different orientations.

regularities are frequently observed (straight domain boundaries parallel to crystallographic planes, near periodicity in a set of alternate domains, etc) (Fig. 2.9).

Each domain is polarized because of the unit cell structure, but the total dipole moment of a set of domains equally distributed between the various possible orientations is zero. Take a crystal slice whose faces are normal to one of the axes, c, and apply an electric field normal to the slice and of increasing strength. The slice becomes polarized, not through the deformation of the structure as in the case of a non-polar crystal but through increasing orientation of the domains into a favourable position in which their moments become aligned more nearly parallel to the field. The favourably oriented domains grow at the expense of others until the crystal as a whole becomes a single domain. The polarization then has its maximum value and reaches saturation. When the applied field is reduced, the crystal at first remains a well-orientated single domain and, at zero field, is still polarized. If the field direction is reversed, domains with inverse polarization begin to grow, and then become enlarged, until with a strong enough field the crystal once again reaches its saturation polarization, this time in the opposite direction: it is once again a single domain. Thus, with an alternating variation in the applied field, the crystal polarization follows a hysteresis cycle which is very similar to that of a ferromagnetic (Fig. 2.10). This explains the origin of the name given to this type of dielectric.

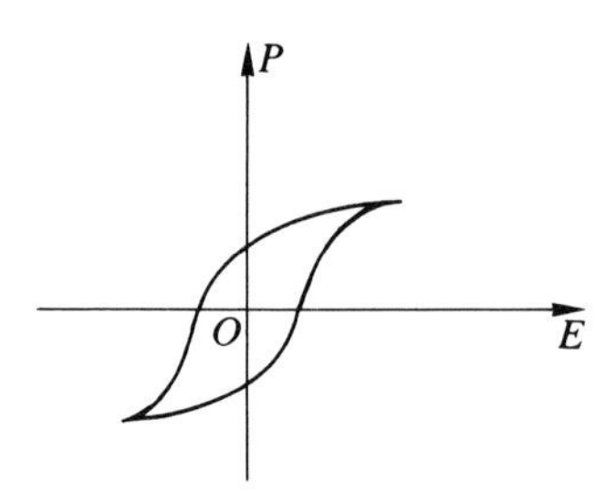

Fig. 2.10. The polarization induced in a ferroelectric by an alternating electric field: a hysteresis loop.

Can 'ferroelectricity' be explained?

How could the properties of ferroelectrics described above be explained? The essential aim here is to provide an explanation of how the structure of the ferroelectric phase comes about. To do this, we have to show that it is the structure with the minimum free energy. However, such a calculation cannot be carried out because we do not know with sufficient accuracy the atomic binding forces occurring in a configuration of given geometry. Here, we encounter one of the main difficulties faced by the physics of solids in connection with many phenomena: some other problems less complicated that than of ferroelectricity still remain unsolved. For example, it is not known how to calculate 'from first principles' the relative permittivity or refractive index of a crystal with a simple structure.

When projects turn out to be too ambitious, physicists abandon them and develop theories described as 'phenomenological'. These start from equations which involve parameters chosen *ad hoc*, or even from models based on empirical data and assumed to have certain properties. Such theories are very useful to specialists in the field of ferroelectricity and enable whole sets of results to be brought together in a coherent whole. However, theories of this sort are not within the scope of the present book, where the aim is to relate properties directly to atomic structure.

A general comment

We have mentioned in this chapter several examples of crystals having a particular property that provides the basis for economically important techniques. The origin of these developments is, in general, an observation made 'by chance' or during an investigation undertaken with a very different aim. There are in practice no miracle materials which have been 'made to measure' according to a recipe drawn up by theoreticians. (Further on, we shall encounter exceptions to this in the field of semiconductors, exceptions that are all the more remarkable for being so rare.)

Nevertheless, the initial discovery cannot be exploited without the help of theoretical physicists in analysing the first observations. Moreover, it is very unlikely that the first sample to be studied will be the one giving the best performance. Hence the need for a wide and systematic search for the best 'candidates', selected for theoretical reasons.

Thus, technical success depends on close collaboration between theoretical and experimental solid state physicists, chemists capable of synthesizing any required materials, and specialists in the field of crystal growth. The last-named are vital since, for many applications, a microcrystalline material is inadequate, and single crystals must be available in large enough sizes, often of the order of a centimetre. It is thanks to collaboration between teams of research workers with varying specialities that the technical wealth from the physics of solids has been, and will in the future be, exploited.

The electrical conductivity of solids

A solid is a conductor of electricity if a current passes through it when a voltage is applied to it. The current is the result of the motion of charged particles or **charge carriers** through the solid.

Ohm's law

We choose the simplest possible geometry: a rod of cross-sectional area S and length L whose ends are maintained at fixed voltages 0 and V by an external source. If there is only one type of carrier, with charge q and density (number per unit volume) n, and if υ is their mean velocity projected along the axis of the rod, the current density is given by:

$$J = I/S = nq\upsilon.$$

Inside the rod, the uniform electric field is $E = V/L$. In the great majority of conductors, the drift velocity of the carriers, υ, is proportional to the electric force on the particle and thus to the electric field, so that $\upsilon = \mu E$, where μ is the carrier **mobility**. Hence:

$$J = nq\mu E.$$

This is an expression of Ohm's law. If we write it in the form $I/S = nq\mu V/L$, we obtain:

$$V = (1/nq\mu)\,(L/S)I = \rho(L/S)I.$$

The ability of a solid to conduct electricity is indicated by the parameter $\sigma = nq\mu$, its **conductivity**, or by its reciprocal $\rho = 1/\sigma = 1/nq\mu$, the **resistivity**.

The charge carrier is subjected to two forces: the electric force qE and an opposing force, the resistance to its motion through the solid. The drift velocity υ has such a value that these two forces are equal in magnitude. This is similar to the way in which a parachute falls with a constant velocity when its weight is exactly balanced by the upward force due to the collisions of air molecules. Ohm's law expresses the fact that the resistance to the motion of the carrier is proportional to the drift velocity.

In order to explain electrical conductivity in terms of a model of atomic structure, the following problems have to be solved: firstly, the carrier or carriers of charge have to be identified; secondly, the causes of the resistance to carrier motion have to be discovered and analysed; thirdly, if Ohm's law is found to be valid, the resistivity must be calculated and compared with experimental values. Finally, we must investigate why some solids do not obey Ohm's law.

The nature of the carriers

There are two types of charged particle capable of transporting electricity through matter: electrons and ions.

Electrons are charged particles, characterized by their charge $-e$ and their inertial mass in the free state m_e. They are point-like, i.e. particles for which the classical concept of a diameter has no meaning.

Ions, i.e. atoms which have lost or gained a small number of electrons, have a positive or negative electric charge of the same order as that of electrons. Their mass is broadly the same as that of atoms, and therefore of the order of a thousand times greater than that of an electron. Finally, they behave as if they had a fairly well-defined diameter (of between 0.1 and 0.5 nm, see SM, p. 9).

Given these properties, it could be predicted that electrons will be the more highly mobile carriers and much more effective than ions, which are too bulky. Electronic conductivity does, in fact, turn out to be predominant and we shall devote most of the chapter to it. Before that, however, we look at a few very special cases of ionic conductivity because of their important applications.

Ionic conduction

The most typical example of this occurs in electrolytic solutions, where the transport of electricity by ions is clearly revealed, at least if we are content with crude approximation. Although liquids fall outside the scope of this book, we look briefly at electrolytic conduction in solutions because it forms a good introduction to the study of the rare cases of ionic conduction in solids.

The motion of ions in ionic solutions produces observable results in that substances are carried to the electrodes and in that there is a relationship between the amount transported and the quantity of electric charge passing through the electrolyte. Faraday's law shows that each ion carries a charge determined by its degree of ionization.

In the absence of an applied electric field, the ions in an electrolytic solution are dispersed among the water molecules. Since these molecules are tightly packed together, the motion of the ions due to thermal agitation is almost always confined to oscillations about fixed points in a cage formed by the surrounding water molecules. However, it may happen, very occasionally, that ions take advantage of a gap to jump into a neighbouring site: this is the effect responsible for normal diffusion in liquids. When an electric field is applied, the ions experience a force qE which tends to move them in the direction of the lines of E. The motion of an ion is, however, opposed by the presence of water molecules, and this is the origin of the resistance that limits ionic velocities, which are very low. They can be derived from a knowledge of the current density and the number of ions per unit volume, and the use of the formula on p. 70. The drift velocity in an electric field of $1\,\mathrm{V\,m^{-1}}$ is of the order of $5\,\mu\mathrm{m\,s^{-1}}$.

From what has just been said, it might be thought that ions of small diameter would have a greater mobility. This is not always what is observed, because matters are complicated by the fact that the ions are not 'bare' but attract water molecules which move along with them. The mobility of a hydrated ion depends on its general shape and size: the more the ion attracts water molecules, the greater its bulk and thus the slower it will move.

Now consider a perfect ionic crystal. When an electric field is applied here, the ions experience the same forces as do the isolated ions in a solution. However, a crystal has a regular structure which is well-defined and held together by strong cohesive forces. This implies that, as soon as an ion is slightly displaced, it will experience a restoring force which quickly exceeds that due to the applied electric field. It is impossible for the ion to escape: the ionic conductivity is zero.

Nevertheless, real crystals do exhibit a certain degree of ionic conductivity, albeit very small and difficult to measure. This is due to the existence of **vacancies** (SM, p. 107), which allow an ion to jump into a similar but unoccupied neighbouring site. The effect is extremely small and ionic crystals are in practice very good insulators.

Covalent and molecular crystals are also very good insulators because, when pure, they contain no free charge carriers.

Dielectric breakdown

A dielectric only remains an insulator if the electric field applied to it is below a certain limit. That is why high voltage equipment has to be fitted with increasingly bulky insulators as the operating voltage increases.

The reason for the existence of a limiting field is that all dielectrics contain a very small number of free electrons, supplied particularly by impurity atoms, which can never be completely removed. In a strong enough field, these electrons are accelerated by it until they have sufficient energy to ionize any atoms they strike. This then liberates further electrons, and so on, until an electron **avalanche** is produced causing an abrupt discharge through the dielectric. The insulator can be permanently damaged by such a discharge.

Another cause of breakdown that should be mentioned is the existence of gas bubbles inside certain ceramics. Under a high voltage, these may act as centres of arc discharges that ruin the insulation.

Ionic crystals with exceptionally high conductivity

In recent years, physicists have become interested in some unusual crystals possessing an abnormally high ionic conductivity. Their resistivity is of the order of $10^2\,\Omega\,\mathrm{m}$, i.e. comparable with that of a semiconductor. We shall describe the special structure that gives rise to this important property by taking the example of β-alumina.

The name of this compound is inappropriate since it is not a form of alumina, but has a composition $11Al_2O_3,1.3Na_2O$. The crystal consists of layers of tetrahedra, formed from one positive Al ion surrounded by four negative O ions joined at their vertices. Between two successive layers of alumina, there are sites that can be occupied by sodium ions. However, the positions of the Na^+ ions are not well defined, because the potential energy minima defining the sites are very shallow: only a small amount of energy is needed to move the ion around its mean position. In addition, there are more sites available than there are ions to be accommodated. It follows that the sodium ions can easily pass from one site to another and can therefore move in the plane lying between the alumina layers. They cannot, however, cross these layers. The crystal is therefore an insulator if the applied field is normal to the alumina planes, but has a high 'two-dimensional' ionic conductivity parallel to the planes.

A collection of randomly orientated microcrystals has a good mean conductivity in all directions, since the conducting planes in adjacent individual crystals have different orientations. β-alumina is an example of a high conductivity ionic crystal.

The interest in β-alumina arises from the fact that it is an essential component in new types of accumulator, light enough to make electric cars possible. In a sodium–sulphur accumulator, the compartments for each electrode must be separated by a sealed wall that is chemically inert and yet offers only a low resistance to the passage of an electric current. β-alumina has these properties, provided its temperature is at least 300 °C.

The electronic conductivity of metals

By far the largest majority of substances that are very good conductors of electricity are metals. Their main characteristic is the lack of any transport of matter accompanying the passage of a current so that it has been natural to think of the electron, ever since its discovery, as being the agent of metallic conduction. Indeed, a simple electron model provided an immediate explanation of the phenomenon and initially met with several successes.

Metals have only a few valence electrons, normally one or two, occasionally three. They are outer electrons, easily detached from their atoms and leaving behind positive ions, which pack closely together and form the framework of the metallic crystal. At the same time, all the free electrons liberated from the atoms form a cloud that envelopes the ions and ensures the overall neutrality of the metal.

The electron cloud is set in motion by an electric field, but it should be stressed that the resultant drift velocity is very small. If we take as an example a copper wire with a cross-sectional area of 1 mm^2 carrying a current of 1 A, the current density is then 10^6 A m^{-2}. The free electron density is equal to the number of atoms per unit volume, i.e. 0.9×10^{29} m^{-3}. Using the formula on p. 70, the drift velocity is calculated to be 0.1 mm s^{-1}, which is insignificant in comparison with the velocities attained by the atoms of the solid during thermal agitation.

While it is carrying the current, the conductor remains neutral: the external voltage source thus provides an electron flux at the negative terminal equal to the electron flux leaving the positive terminal.

Many problems have to be solved if we are to arrive at a model yielding quantitative results. How can the resistivity of a given metal be predicted? Why does it vary from one metal to another when the number of free electrons is the same and the crystal structures are very similar? How can the variation of conductivity with temperature be explained? Why do impurities, even in very small proportions, have so much influence on the resistivity of a metal? And these are only some of the questions to be raised.

Our understanding of conduction in metals gradually deepened through the use of models that were increasingly refined and adapted. However, it was the introduction of quantum ideas that produced the greatest advances in the theory of an apparently very simple phenomenon.

The classical theory of electrons in metals

In this theory, due to Drude and Lorentz (1905), it is assumed that the charge carriers are free electrons. The initial problem is to discover why there is any resistance to the drift velocity of the electron cloud. Our starting point is a representation of the cloud using a model inspired by that of a perfect gas.

The motion of the electrons due to thermal agitation is completely random and they do not interact with each other. They are also free, although they move around in the electric field produced by the positive ions and the other electrons. Their paths are broken up by collisions with the other electrons or the ions. All these conditions are very similar to those governing molecular motion in a perfect gas.

For a given metal at a given temperature, there is an average time taken for the free path between two successive collisions. We simply assume that these times are all equal to τ; the mean free path has a constant length corresponding to this time τ. Such a model may seem very crude but, with the degree of approximation we are adopting, it would not be particularly useful to complicate it further.

While travelling over its free path, the electron experiences the electric force $-e\boldsymbol{E}$ and acquires a component of velocity along the direction of $\boldsymbol{E}$

which varies linearly from 0 to $\upsilon = eE\tau/m_e$. The subsequent collision sends the electron in a random direction, so that it is made to lose the whole of its velocity component parallel to $\boldsymbol{E}$. During the next free path, this velocity component once again increases from 0 to υ. It is therefore as if the electron had a constant drift velocity of $\upsilon/2 = eE\tau/2m_e$. If n is the number of free electrons per unit volume, the current density is:

$$J = ne\upsilon/2 = ne^2E\tau/2m_e.$$

The conductivity is therefore $\sigma = J/E = ne^2\tau/2m_e$. Apart from the universal constants e and m_e, and the easily determined free electron density n, a calculation of σ involves only the single parameter τ. Conversely, a value for τ can be deduced from an experimental value of σ. Thus for copper at ordinary temperatures, we find that $\tau = 2 \times 10^{-14}$ s.

Furthermore, if the perfect gas model is applied to the electron cloud, so that the 'point charges' play the role of the gas molecules, the mean kinetic energy of the electrons is $3k_BT/2$ and at 300 K this corresponds to a velocity of $10^5\,\text{m s}^{-1}$, which is enormous compared with the drift velocity of about a millimetre per second due to the applied electric field. From the mean speed u due to thermal agitation, it is possible to calculate the mean free path Λ, which is the distance travelled during the time τ. Thus $\Lambda = u\tau$, or $\sigma = ne^2\Lambda/2m_eu$. In our example of copper at room temperature, this gives $\Lambda = 2$ nm.

This value is a little high, since there is one electron per unit cell of side 0.36 nm, but it is not all that improbable. Conversely, it could be said that if a reasonable value for Λ is assumed, classical theory yields a value for the resistivity of copper of a satisfactory order of magnitude, which counts as a success for such a simple theory.

On the other hand, serious objections can be raised against the theory in view of the very unlikely consequences that follow from it.

1. From the expression adopted for the mean kinetic energy of agitation of the free electrons, $\frac{1}{2}m_eu^2$, their mean speed u is predicted to be proportional to $T^{1/2}$. If the mean free path is independent of the temperature, or nearly so, it follows that the resistivity $\rho = 1/\sigma$ should also vary as $T^{1/2}$. Now to a good approximation, it is observed that ρ varies linearly with T over a wide range of temperatures.

2. If the mean kinetic energy of a free electron is $3k_BT/2$, a mole will have a heat capacity of $3R/2$ due to the free electrons (one electron per atom), which is to be added to the heat capacity of the lattice equal to $3R$ (p. 3). This means that for a monovalent metal the molar heat capacity should tend to $9R/2$ at high temperatures, which is not what is observed: the heat capacity of a metal is only slightly greater than $3R$, as in the case of insulators. The electrons apparently have no heat capacity, and this is a fact that is inconsistent with the classical model of the electron gas.

In spite of these failures, credit must nevertheless be given to the Drude–Lorentz theory for a correct prediction of a relationship between the electric and thermal conductivities of metals. It is well known that metals are good conductors both of electricity and heat and in both cases the active agents are the free electrons. We have already established a value for the thermal conductivity due to particles using the kinetic theory of gases. According to the formula on p. 54, and using the fact that the heat capacity per unit volume for the electron gas is $3nk_B/2$, we can write:

$$K = \frac{1}{3} \cdot \frac{3}{2} nk_B u\Lambda.$$

Dividing this equation by that for the electric conductivity σ on page 75, we find that:

$$\frac{K}{\sigma} = \frac{1}{2} \frac{k_B u^2 m_e}{e^2}$$

Since $\frac{1}{2}m_e u^2 = 3k_B T/2$, we end up with

$$\frac{K}{\sigma T} = 3\left(\frac{k_B}{e}\right)^2.$$

This therefore predicts that the ratio $K/\sigma T$ is a *constant* which is, moreover, independent of the nature of the metal since forming the quotient K/σ removes the only parameter, Λ, which characterizes the metal. In fact, Wiedemann and Franz established empirically in 1853 that the ratio $K/\sigma T$ was close to 2.5×10^{-8} W Ω K^{-2} for many metals. The extremely rough calculation we have made leads to a value of 2.2×10^{-8} W Ω K^{-2}.

What can be deduced from these various comparisons between theory and experiment? We have here a very common and typical situation. Disagreement with experiment invalidates a theory, whereas agreement does not constitute a certain proof of its validity, since it may be the result of chance or of interactions that have not been appreciated as relevant.

Thus, it cannot be said that the Wiedemann–Franz law provides justification for the Drude theory. In fact, this theory is not a valid one for evaluating either the thermal or the electric conductivity on their own. What has happened is that the inaccuracies in the theory have fortunately cancelled each other out in calculating the ratio of the two conductivities.

Drude's theory predicts a parabolic increase in resistivity with temperature, while experiment shows a linear increase. Whatever the coefficients describing the two curves, there is always one value of temperature at which they intersect (Fig. 2.11). Thus, in every case, there is a temperature at which Drude's theory gives a result close to that obtained by experiment. This is what happens by chance for a very common metal, copper, near room temperature.

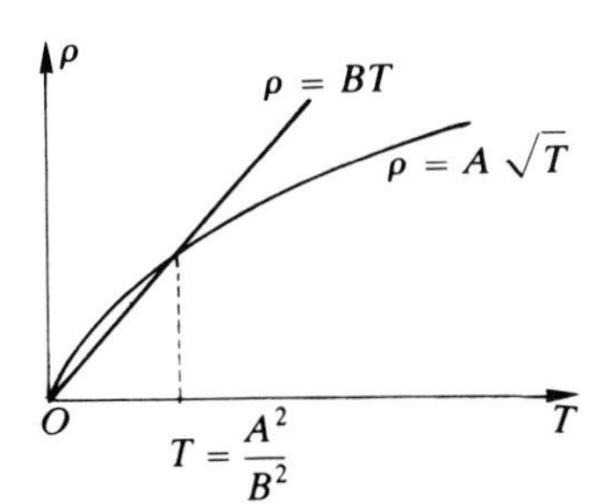

Fig. 2.11. Fortuitous agreement between the Drude theory and experiment: whatever the values of the coefficients A and B, there is one temperature at which the predicted and observed values of the resistivity agree.

Quantum theory of electrons in metals

Quantization of energy had to be introduced into the treatment of the heat capacity of solids to account for its falling to zero at the absolute zero. Until now, however, in this book as in the other (SM) dealing with structures, classical atomic models have been adequate since it has not been necessary to go as far as the subatomic scale. On the other hand, in order to relate the behaviour of electrons in metals to experimental results on conductivity, classical physics is no longer adequate: we have to call on quantum mechanics.

To be logical and rigorous, our account should start from the basic equations and postulates of quantum mechanics, from which the ideas describing the behaviour of electrons are derived. We shall adopt a different approach: the quantum ideas will be 'stepping stones', assumed without proof, and we shall use them to explain the results, often without detailed calculation. One of the aims in this book is to familiarize those who cannot assimilate mathematical complexities with the results of modern physics: the semiconductor technologist, for instance, has to use quantum principles in everyday work without the need to use any complicated theoretical calculations.

Two successive quantum mechanical models have been used to represent conduction electrons in solids: the first, very crude, model is nevertheless important, since it removes the serious problem of the absence of a heat capacity due to the electrons in metals. The second, more realistic, model leads to a new way of looking at insulators and conductors and opens up a very rich intermediate area, that of semiconductors.

Electrons in an empty box

We start from the idea that 'free' electrons can move around inside a piece of metal but cannot get outside it. The electron moves in an electric field created by the positive ions located at the nodes of the crystal lattice. Around each of these nodes, the electron is attracted by the positive charge which is surrounded by an attractive potential well. Note, in addition, that the potential due to the ions is modified by that due to the electrons, so that the situation is very complicated.

The first approximation, made by Sommerfeld (1928), is to assume that the electrons inside the metal are moving in a *uniform electrostatic potential*, which is positive with respect to the potential in the space outside. The potential barrier (Fig. 2.12) thus confines the electron to the interior. Not only does this model neglect the spatial variations in potential due to the ions, but it also neglects the electric interactions between electrons. This is the **empty box** model, in which the only variation in potential that occurs is due to the metal surface. The whole assembly of free electrons, with a known density, is moving within this box.

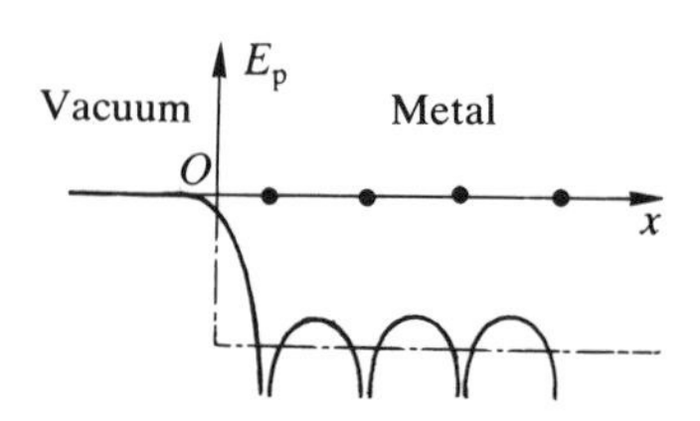

Fig. 2.12. Principal features of the variation in potential energy of an electron in a metal along a row of ions. The broken line indicates the value used in the first approximation.

First step: one electron in the box

The first problem to be solved is how to describe the state of a single electron in the box. Consider first an electron in an infinite space at constant potential, and therefore moving in a straight line with a constant velocity. In classical physics, this is represented by a point mass m_e with a well-defined position at a given instant and with a momentum $\boldsymbol{p} = m_e\boldsymbol{v}$. Its kinetic energy is:

$$E = \tfrac{1}{2}m_e v^2 = p^2/2m_e.$$

In quantum mechanics, the electron in uniform straight line motion is associated with a plane wave of wavelength λ, which is generally defined by its wave vector $\boldsymbol{k}$ in the same direction as the velocity of the particle and having a magnitude $k = 2\pi/\lambda$. The position of the particle is indeterminate since the wave fills the whole of the space.

The correspondence between the two representations is made through the de Broglie relationship:

$$\lambda = h/p \quad \text{or} \quad \boldsymbol{p} = \hbar\boldsymbol{k}.$$

The wave vector $\boldsymbol{k}$ is thus *associated* with an electron of kinetic energy:

$$E = \hbar^2k^2/2m_e.$$

The free electron must now be 'caged' in a box. The fact that it cannot escape is expressed by saying that the wave is reflected by the walls. We start by taking a one-dimensional case in which an electron of momentum p, when confined to a segment of a straight line, oscillates between two fixed points a distance L apart. The associated wave, a sinusoidal wave with one variable, is reflected at both ends of the segment.

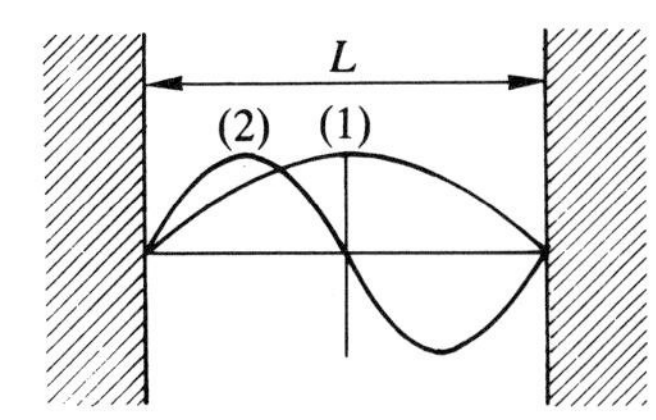

Fig. 2.13. Two normal modes of the wave associated with the electron confined to the line segment L corresponding to the wave vectors $k_1 = \pi/L$ and $k_2 = 2\pi/L$.

We recognize here the well known problem of a vibrating string: there are **normal modes** whose wavelength is an integral fraction of the length $2L$, i.e. $2L/n$, or in other words whose wave vector is an *integral multiple* of π/L, i.e. $k_n = n\pi/L$ (Fig. 2.13). In the same way, the electron in the segment L is found to be in a **proper state** or **eigenstate** characterized by a wave vector k_n belonging to the arithmetic series $k_n = n\pi/L$. The possible electron states correspond to the *discrete* set of energies:

$$E_n = n^2 \frac{\hbar^2\pi^2}{2L^2 m_e}.$$

This result can be generalized to the electron confined in a box, and we then find that we recover what has already been calculated in connection with the normal modes of the elastic vibrations in a solid (p. 25). We simply reproduce the conclusions from the analysis in Box 5. If we take a cubic box of side L, the wave vector $\boldsymbol{k}$ corresponding to a possible state of the electron in the box has components along the three directions parallel to the edges of the cube equal to $n_1\pi/L$, $n_2\pi/L$, $n_3\pi/L$, where n_1, n_2, n_3 are integers.

Box 8

Density of states for an electron in a box

The problem of an electron confined to a cubic box defined by

$$0 < x < L, \quad 0 < y < L, \quad 0 < z < L$$

can be reduced to a study of the associated wave. As we saw in Box 5, because of the reflections at the outer surfaces of the box, the wave must take the form:

$$A = A_0 \sin(k_x x) \sin(k_y y) \sin(k_z z) \exp(i\omega t)$$

with $$k_x = n_x\pi/L,\ k_y = n_y\pi/L,\ k_z = n_z\pi/L,$$

where n_x, n_y, n_z are positive integers. Following the same argument as in Box 5, we find that the number of wave vectors with magnitudes lying between k and $k + \Delta k$ is

$$Vk^2\Delta k/2\pi^2$$

where $V = L^3$ is the volume of the box.

The present problem involves waves associated with an electron, which are scalar waves and not vector waves as they are in the case of lattice vibrations. Because of the spin $\frac{1}{2}$ of the electron, we assume that there can be two possible states for each point $\boldsymbol{k}$ in $\boldsymbol{k}$ space. The number of states corresponding to wave vectors with magnitudes lying between k and $k + \Delta k$ is therefore:

$$Vk^2\Delta k/\pi^2.$$

The only factor involved in this formula is the volume of the box and, although it has been derived for the case of a cube, it turns out to be valid for any shape.

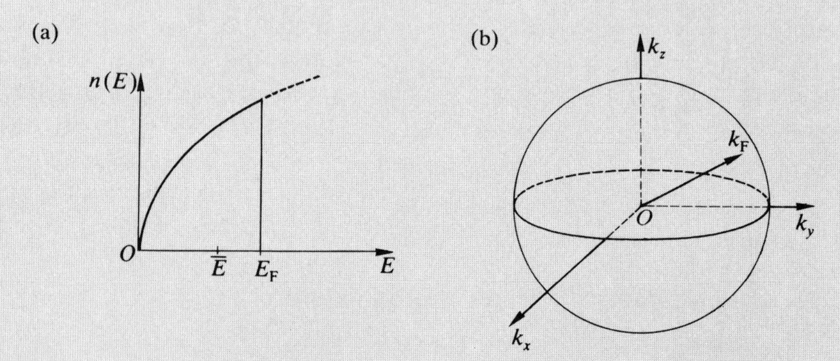

Fig. 2.14. (a) Density of occupied states at $T = 0\,\mathrm{K}$ and the upper limit at the Fermi level, E_F; (b) the distribution of the wave vectors of occupied states in the Fermi sphere of radius k_F.

For free electrons, we know that the relationship between the modulus of the wave vector $\boldsymbol{k}$ and the kinetic energy of the electron is:

$$E = \hbar^2 k^2 / 2m_e.$$

By making a simple change of variable, we derive from this the number of electron states with energy between E and $E + \Delta E$, denoted by $n(E)\Delta E$:

$$n(E)\Delta E = \frac{V}{2\pi^2}\left(\frac{2m_e}{\hbar^2}\right)^{\frac{3}{2}} \sqrt{E}\,\Delta E.$$

The density of states $n(E)$ thus varies as the square root of E in the case of free electrons (Fig. 2.14).

These results call for several comments.

It is a characteristic of quantum theory to predict discrete states of constant energy for a system. One of the first successes of the new mechanics, it will be recalled, was accounting for the electron states in the hydrogen atom which explained the observation of sharp spectral lines.

Although this system of the electron in a box is a simpler one, discrete states again occur. However, the differences between the wave vectors, or between the energies of adjacent states, are extremely small because of the large ratio of the macroscopic magnitude L to interatomic distances: because of this, it might be thought that no significance should be attached to the fact that the variation of the vector is discontinuous. However, we shall see in the next section that the indirect consequences are fundamental to the behaviour of electrons.

As we have already mentioned in connection with the elastic mode of vibration, it seems curious that the scale of the energy states is determined by an external magnitude which has no connection with the intrinsic properties of metal. In fact, the quantities V and L are simple intermediate parameters in the calculation and they do not appear in the evaluation of quantities capable of being compared with experiment. Neither the total number of free electrons nor the volume of the metal are separately involved, but only the number of electrons per unit volume, or the electron density, which is an intrinsic property of the metal.

Second step: N electrons in the box

We shall neglect any electrostatic interaction between electrons. This is again an approximation: in classical language, it means that the 'electron gas' is considered to be a perfect gas and, in quantum language, it means that the electron waves are not coupled. The possible states for each of the N electrons are therefore those which were defined for the isolated electron in the same box.

However, in quantum theory there is an additional constraint on electrons that did not exist in the case of lattice vibrations. This is the **Pauli exclusion principle**, which can be stated as follows: a single energy level can only be occupied by one or two electrons. If there are two electrons with the same wave vector, they must have opposite spins, i.e. if a magnetic field is applied, the magnetic moments of the two electrons align themselves in opposite directions.

The Pauli principle applied to the electrons in an atom provides the basic explanation for the building up of electron layers or shells in the atom and thus for the Mendeleev periodic classification of the elements. The principle also applies to the electrons confined in a single box since the wave associated with each of them is to be found over the whole of the space within the box.

Because only two electrons can be accommodated in each energy level, the accommodation of N electrons must mean that at least $N/2$ levels are occupied. Consider first the situation at absolute zero: the equilibrium state of the system corresponds to that in which the total energy of all the electrons is a minimum. Starting from the lowest level, states with increasing energy must be occupied at the rate of two electrons per level. The level corresponding to the state with the maximum energy is known as the **Fermi level** and its energy is denoted by E_F. The associated wave vector is the **Fermi wave vector**, $\boldsymbol{k}_F$. The waves associated with the occupied electronic states thus have a wave vector $\boldsymbol{k}$ with a magnitude less than k_F. In k-space, the ends of the vectors $\boldsymbol{k}$ are contained within a sphere of radius k_F: this is the **Fermi sphere** (Fig. 2.14).

Box 9 shows how to calculate the Fermi wave vector in terms of the electron density and the mean energy of the N electrons. The calculation gives maximum energies of the order of 8 eV and a mean energy of around 5 eV.

Box 9

Calculation of the Fermi level and the mean energy of free electrons

In Box 8, we calculated the number of electronic states with energies between E and $E + \Delta E$. At 0 K, the N electrons are distributed at the rate of one per state over the states with energies ranging from $E = 0$ to $E = E_F$. The value of E_F, the Fermi energy, is calculated by expressing the fact that the total number of occupied states between E and E_F is indeed equal to N:

$$N = \int_0^{E_F} n(E)\mathrm{d}E.$$

(Note that $n(E)$ contains the factor 2 relating to the two possible spin states.)

Using the expression for $n(E)$, we obtain:

$$N = \frac{V}{3\pi^2}\left(\frac{2m_e E_F}{\hbar^2}\right)^{\frac{3}{2}}$$

and hence

$$E_F = \frac{\hbar^2}{2m_e}(3\pi^2 n)^{\frac{2}{3}}$$

in which $n = N/V$ is the electron density.

From the relationship $E_F = (\hbar^2/2m_e)k_F{}^2$, the Fermi wave vector is obtained as:

$$k_F = (3\pi^2 n)^{1/3}.$$

The mean kinetic energy of an electron can also be calculated using the following expression:

$$\bar{E} = \frac{1}{N}\int_0^{E_F} En(E)\mathrm{d}E.$$

Thus

$$\bar{E} = \frac{1}{N}\frac{V}{5\pi^2}\left(\frac{2m_e}{\hbar^2}\right)^{\frac{3}{2}} E_F{}^{\frac{5}{2}},$$

or

$$\bar{E} = 3E_F/5.$$

We now calculate an order of magnitude for $\bar{E}$ and E_F. There is one electron per atom in a monovalent metal. That implies an electron density of the order of $10^{29}\,\mathrm{m}^{-3}$ assuming an interatomic distance of a few tenths of a nanometre. Using the values for $\hbar$ ($1.05 \times 10^{-34}\,\mathrm{J\,s}$) and the electron mass ($m_e = 9 \times 10^{-31}\,\mathrm{kg}$), we obtain:

$$E_F = 12.6 \times 10^{-19}\,\mathrm{J} \text{ or } 7.9\,\mathrm{eV}$$

and

$$\bar{E} = 4.7\,\mathrm{eV}.$$

(The electron volt is a very appropriate unit for the orders of magnitude of kinetic energies of electrons in metals.)

The Sommerfeld quantum model, simple as it is, leads to results of considerable significance since they are in complete contradiction with classical predictions.

In classical theory, the mean energy of a free electron is $3k_BT/2$, and is therefore zero at 0 K. We now find that it is equal to the energy that a classical electron would have at a temperature of 50 000 K. At the Fermi level, the electrons have an energy that, in classical theory, is only reached at 80 000 K. The most energetic electrons have a speed of $10^6\,\mathrm{m\,s^{-1}}$ at 0 K.

If this is so, it is because of the Pauli principle: the N electrons cannot accumulate in the lowest energy levels. Since these levels are discrete and their energy increases as the square of k, the N electrons can only be accommodated by using levels with high energy. We shall see that these statements are confirmed by the facts.

The effect of temperature on electron energies

The distribution at absolute zero is modified at any higher temperature because the electrons are excited. Some leave their level to jump into one having a greater energy and with an unoccupied state available. However, even at a really high temperature, say 1000 K, the energy due to thermal excitation of the electron, about k_BT, is very small in comparison with the Fermi energy. Electrons inside the Fermi sphere are therefore unable to reach an empty level; only the states represented by a point located in the sphere's 'skin' of thickness k_BT have an appreciable chance of being promoted.

As a result of this, the theory predicts that the Fermi sphere above 0 K, instead of having a sharp boundary, is terminated by a crust in which the population of levels is less than 2 and tends continuously to zero (Fig. 2.15). The thickness of the crust in terms of energy is k_BT. The empty states left below E_F are compensated by occupied states above E_F.

It follows that the mean electron energy is greater than that at 0 K, but not by very much. This is because the number of 'promoted' electrons is relatively small: qualitatively, it can be seen that the number is proportional to the thickness of the sphere's 'crust', and thus to k_BT. Moreover, each promoted electron gains energy of the order of k_BT and the increase in the mean electron energy must therefore be proportional to the square of the temperature.

The corresponding heat capacity per mole of free electrons is $C = \gamma T$. Box 10 shows how to calculate an expression for the parameter γ. The main

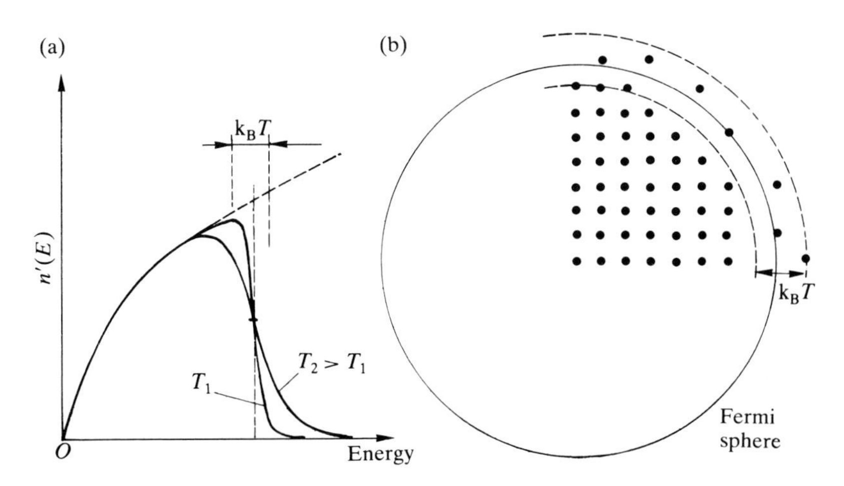

Fig. 2.15. (a) The density of occupied states $n'(E)$ at two different temperatures; (b) the distribution of the wave vectors of occupied states in the Fermi sphere at $T > 0$.

point is that γT is very small compared with $3R/2$, which would be the heat capacity of the electron gas if it were treated classically.

Box 10

The Fermi–Dirac distribution and the electronic heat capacity

The effect of temperature on the distribution of electrons over the various electronic states is described quantitatively by the Fermi–Dirac function, which gives the probability that a state with energy E is actually occupied by an electron at temperature T:

$$f(E) = \frac{1}{1 + \exp \dfrac{E - E_F}{k_B T}}$$

This function is plotted in Fig. 2.16. It can be seen that, for $T = 0$, it is a step function: $f(E) = 1$ for $E < E_F$ and $f(E) = 0$ for $E > E_F$. This means that all the states are filled up to $E = E_F$, while all the states with higher energies are empty.

Above 0 K, the number of electrons with energies between E and $E + \Delta E$, $n'(E)\Delta E$, is equal to the product of the number of possible states $n(E)\Delta E$ in this energy range and the probability of occupation

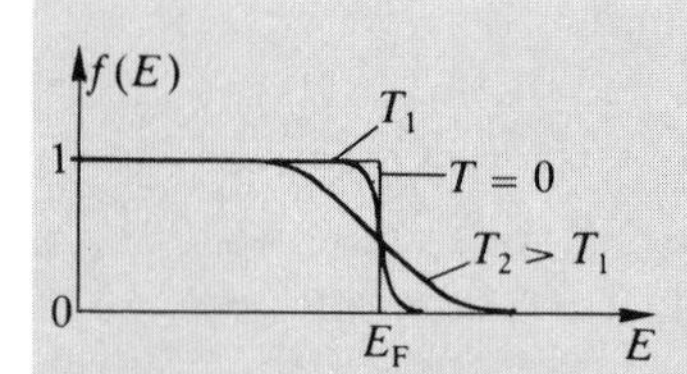

Fig. 2.16. The Fermi–Dirac function: probabilities of occupation of a state with energy E at temperatures $T = 0$, T_1 and $T_2 > T_1$.

of the state of energy E given by the Fermi function. This is how the distributions of electrons at various temperatures shown in Fig. 2.15(a) are obtained.

However, as can be seen in Fig. 2.16, because of the exponential form of the denominator of $f(E)$, the effect of temperature only makes itself felt in a 'band' of energy of thickness $k_B T$ around $E = E_F$. Out of the N_A electrons in a mole of the metal, a certain number N_1 gain an additional energy of the order of $k_B T$. At ordinary temperatures (below 1000 K), this energy is much less than E_F. When the temperature rises from 0 K to T, the total gain in energy can be estimated to be:

$$\Delta E = N_1 k_B T.$$

The N_1 electrons involved occur in the band of energy of thickness $k_B T$ about the Fermi energy,

$$N_1 = n(E_F) k_B T.$$

The increase in energy of the system is therefore:

$$\Delta E = n(E_F) {k_B}^2 T^2.$$

The heat capacity is obtained by taking the derivative of ΔE with respect to temperature:

$$C = 2n(E_F) {k_B}^2 T.$$

It is found that this approximate argument gives a good order of magnitude for the electronic heat capacity. An exact calculation can only be carried out using what are known as 'Sommerfeld' integrals and taking into account the variation of the Fermi level with temperature. However, this only changes the coefficient and the exact calculation gives:

$$C = \frac{\pi^2}{3} n(E_F)\, {k_B}^2 T.$$

It can be seen that a knowledge of the electronic heat capacity enables the density of states of the Fermi level to be assessed. It is remarkable that we have assumed nothing about the form taken by the density of states. This expression is therefore very general and is valid outside the approximations of free electron theory. If we now restrict ourselves to the limitations of this theory, we can use the expression for $n(E_F)$ given in Box 8:

$$n(E_F) = \frac{V}{2\pi^2}\left(\frac{2m_e}{\hbar^2}\right)^{\frac{3}{2}} E_F^{\frac{1}{2}}$$

together with the relationship between E_F and N_A, the number of electrons per mole (Box 9):

$$N_A = \frac{V}{3\pi^2}\left(\frac{2m_eE_F}{\hbar^2}\right)^{\frac{3}{2}}$$

We then arrive at the simple expression:

$$n(E_F) = 3N_A/2E_F,$$

which enables us to rewrite C in the form:

$$C = \frac{\pi^2}{2} R \frac{k_BT}{E_F}$$

where we have used $R = N_Ak_B$.

This last expression enables us to obtain some idea of the order of magnitude of C. The ratio k_BT/E_F is no larger than a few hundredths at ordinary temperatures, and this means that the electronic heat capacity, although it increases with temperature, always remains much less than the contribution from lattice vibrations (which, by Dulong and Petit's law, is of the order of $3R$).

Experiment provides good verification of the quantum calculation. Although the electronic heat capacity is in general very small compared with

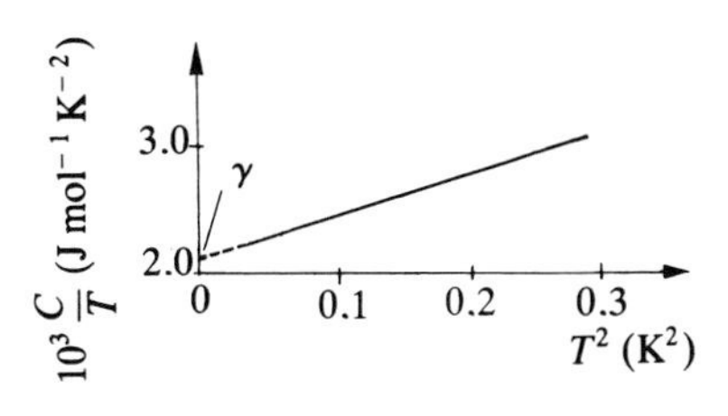

Fig. 2.17. Molar heat capacity of potassium at a temperature near $T = 0$ K. The term due to phonons, AT^3, is supplemented by the term due to free electrons, γT: $C = AT^3 + \gamma T$ or $C/T = \gamma + AT^2$.

the heat capacity due to lattice vibrations, this is not so at very low temperatures. This is because the heat capacity due to phonons (p. 28) decreases to zero at T^3, while that due to electrons decreases linearly with T and below 1 K becomes predominant. Accurate measurements of heat capacity thus give an accurate value for γ (Fig. 2.17). The experimental values have the right order of magnitude (from 0.5 to $5 \times 10^{-3}\,\mathrm{J\,mol^{-1}\,K^{-2}}$) but the agreement with the theoretical expression is not very good since the empty box model is oversimplified.

The almost-free electron model: the introduction of band theory

Although the first model of perfectly free electrons in a box has yielded some very important results, it is obvious that it cannot explain the properties of a particular crystal since none of its characteristics are taken into account.

To proceed any further, it is necessary to introduce the fact that the electrons are moving in a region where the potential is varying because of the positive ions present. The main point is that, with the ions being arranged regularly in the perfect crystal, the potential they produce is periodic with the same periods as the crystal lattice. Apart from that, we retain the approximations already made in the case of the empty box: i.e. we examine the behaviour of an isolated electron without taking into account the interactions between electrons.

The quantum view is considerably different from classical ideas. According to the latter, the point-like mass is imagined to be forcing a passage through the metal ions, its motion being impeded by the collisions it suffers. Now, however, our model of the metal is an empty box in which the potential created by the ions is modulated, with a strict periodicity if the crystal is perfect. The wave associated with the electron is propagated inside the metal without decay or deformation *in spite of* the presence of ions.

The difference between this model and that of the box at a uniform potential is that the wave function of the electron is no longer a single sinusoidal wave but has a spatial structure with the periodicity of the crystal. What that means is this: at constant potential, the electron is completely unlocalized, i.e. the probability that it is present is the same at all points in the space traversed by the wave. In the crystal, on the other hand, the wave amplitude is modulated at the same period as the lattice and the probability of finding the electron near the ions is greater than that of finding it some distance away from them.

As in the first model, an electronic state is specified by its wave vector $\boldsymbol{k}$ and corresponds to a certain energy. In the empty box, this had the value $\hbar^2k^2/2m_e$ for any value of $\boldsymbol{k}$, but this is not always the case in the metal crystal. The difference is in fact small for small values of k: an electron with a long wavelength is 'almost free'. The effect of the modulation of the potential only makes itself felt when the wavelength becomes close to the interatomic distance.

In this case, we revert to a situation already mentioned for phonons (p. 23). Elastic waves of long wavelength are propagated through the crystal almost as they are in a continuous medium of uniform density, and significant differences only appear when the wavelength is of the same order as the interatomic distances.

The interaction between the electrons and the crystalline material is revealed by the phenomenon of **electron diffraction**. Consider a thin crystal slice cut parallel to a family of lattice planes (Fig. 2.18). A monoenergetic beam of electrons is directed perpendicularly on to this slice, the electrons being associated with a wave having a fixed wave vector determined by the energy. The electrons pass through the slice and are thus propagated freely through it, except when their wavelength is equal to twice the interplanar distance, in which case the electrons are reflected by the slice. This will be recognized as a special case of **Bragg reflection** of the waves by the crystal. We mention this experiment because it shows the reality of critical situations for the interaction of electrons and a crystal.

Fig. 2.18. Reflection of an electron beam directed along the normal to a family of lattice planes: a special case of the Bragg condition.

Before beginning the study of the propagation of electron waves in a crystal, we look once more at the simpler example of a one-dimensional system: electrons are propagated along a segment of a finite straight line of length L. Along the axis, the potential varies periodically with the period a of the positive ions arranged along the line segment. If the modulation of the potential produced by the ions is small, the energy $E(k)$ is not very different from that predicted by the free electron theory. The first critical values of k are $\pm\ \pi/a$ (i.e. $\lambda = 2a$). Near these values, the energy of the state is different from the free electron value, becoming smaller below π/a and higher above π/a. For $k = \pi/a$, there are *two* levels with energies separated by a gap (the forbidden band). These two values are the limits of the energy curve at each side of the critical point at $k = \pi/a$. Other critical values of k occur at integral multiples of π/a, $k = n\pi/a$ (Fig. 2.19).

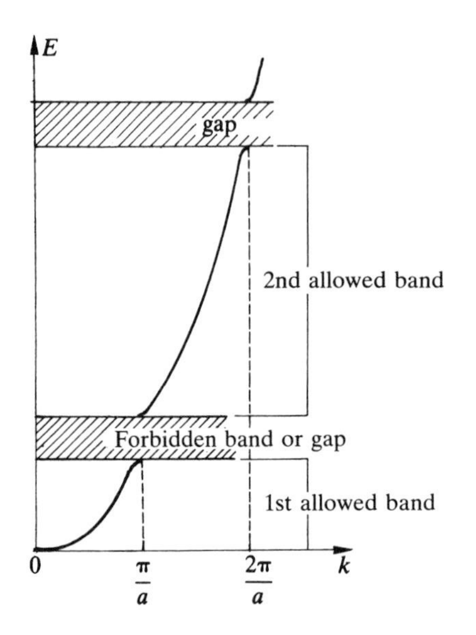

Fig. 2.19. Energy bands of electrons in metals.

All these values correspond to a solution of the **Bragg equation** adapted to the one-dimensional case, $n\lambda = 2a$. The reflection of the electron wave at critical values of the wave vector occurs because the propagation of the wave is considerably modified. Quantum mechanics shows that, when the wave vector is close to the critical values, the interaction of the wave with matter is always strong enough to produce a considerable change in the energy of the free electron.

This example, oversimplified as it is, reveals two results which are important because they can be generalized to the real case of a metal crystal.

1. Looking at the energies of possible states, a series of **bands** appears, within each of which the discrete states succeed each other very closely. Two successive bands are separated by a **forbidden band**: there are no possible states corresponding to an energy lying in a forbidden band. This set of gaps in the possible states is caused by the periodic modulation of the potential.

2. Because the one-dimensional lattice is finite, the values of k corresponding to possible states are discrete and regularly spaced. The range of k values lies between $-\pi/a$ and π/a, and thus corresponds to what we have called the Brillouin zone (see Box 4). In that case, we showed that the

number of states in k space is equal to the number of ions N in the line segment. The number of energy levels in a band is therefore N. Since, according to the Pauli principle, there can be at most two electrons per level, the first band can contain a maximum of $2N$ electrons. This result is expressed in a specific form related only to the atomic structure of the object: the number of levels in the first band is one per unit cell of the lattice, which corresponds to a maximum of two electronic states per unit cell.

We now move on to the real case, that of a three-dimensional metal crystal. The waves associated with the electrons are propagated in any direction, so that their wave vectors specify both the value of the wavelength and the direction of propagation. Again, we revert to a situation already described for elastic waves.

The critical values of $\boldsymbol{k}$ for which there are discontinuities in the electron energies are related to the reciprocal lattice of the metal crystal via the Brillouin zones (p. 19). The perpendicular planes (median planes) bisecting all the reciprocal lattice vectors starting from the origin are drawn. The nearest of these define a multi-faceted polyhedron centred on the origin which we have called the **Brillouin zone**. It would be more correct to call this the first Brillouin zone, since the median planes of all the reciprocal lattice nodes define volumes which are embedded inside each other, forming Brillouin zones of increasing order.

The electron energy for small k values (at the centre of the first zone) is the same as that for an electron in an empty box, $\hbar^2k^2/2m_e$. As $\boldsymbol{k}$ approaches the zone boundaries, the energy begins to differ from that of the free electron. At the zone surface, there is a discontinuity (the energy gap), and the energy then increases again in the next and successive zones. These discontinuities at the zone boundaries are related to the fact that,

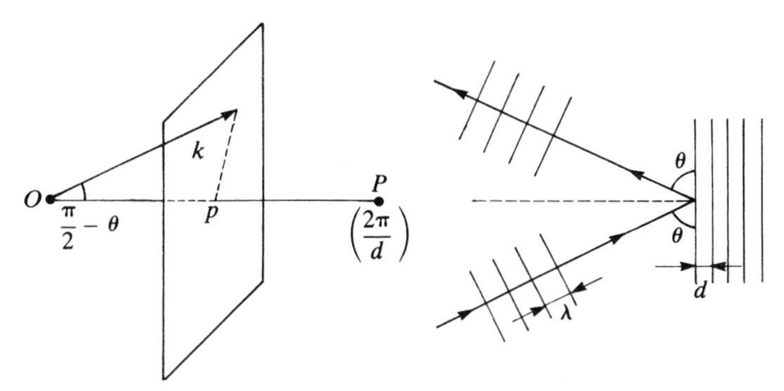

Fig. 2.20. The condition for Bragg reflection is satisfied by the wave associated with an electron when its wave vector reaches the boundary of the Brillouin zone.

when the vector $\boldsymbol{k}$ reaches the surface of a zone, the associated wave experiences Bragg reflection at a family of lattice planes (see Box 11).

Box 11

Bragg reflection of electron waves

Let P be a node of the reciprocal lattice of the metal crystal (Fig. 2.20). The length OP is $2\pi/d$, where d is the interplanar spacing normal to the direction $\overrightarrow{OP}$. The vector $\boldsymbol{k}$ has a length $2\pi/\lambda$ and makes an angle $\pi/2 - \theta$ with $\overrightarrow{OP}$: if it reaches the median plane of the pair of points $[O, P]$, it has a projection p on to OP such that $\overrightarrow{Op} = \frac{1}{2}\overrightarrow{OP}$, i.e.

$$\frac{2\pi}{\lambda}\cos\left(\frac{\pi}{2} - \theta\right) = \frac{1}{2}\frac{2\pi}{d}.$$

This relationship is exactly the same as the Bragg equation applied to waves of wavelength λ propagated along the direction of $\boldsymbol{k}$ and reflected by lattice planes of interplanar spacing d normal to $\overrightarrow{OP}$:

$$\lambda = 2d\sin\theta.$$

If the states are classified in groups of increasing energy, it may happen that there is a gap between the maximum energy of a state in the first zone and the energy of the states in the second zone. We find in the metal, as in the one-dimensional model, that there is once again a forbidden band separating two allowed bands. However, the case of a crystal is more complicated since it is essentially anisotropic. The energy values depend on the direction of the vector $\boldsymbol{k}$: it is conceivable that overlapping of the energy bands of states whose wave vectors have different directions may

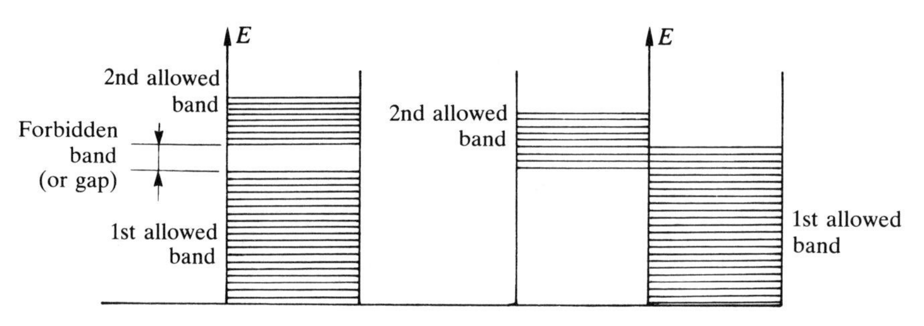

Fig. 2.21. Energy bands with forbidden band or gap and overlapping bands.

mean that no gaps are left in the energy range covered by all the electrons (Fig. 2.21).

As regards the number of states in the first Brillouin zone, the situation is simple: whatever the crystal structure, the number of levels is equal to the number of crystal unit cells contained in the metal block; the first zone can contain at most two occupied states per unit cell, each level being occupied by two electrons with opposite spins.

Free electrons and bound electrons in the metal

In the model just described, the valence electron or electrons were clearly distinguished from those in the other atomic shells. While the valence electrons are almost free in the metal, it is assumed that the states in the inner shells are the same as those in the isolated atom.

In reality, the representation of an electron by a wave traversing the whole crystal is valid for all the electrons. The consequence of this is that for all of them the precisely defined level in the free atom is replaced by a set of discrete levels distributed over a band of energies. For the inner shells, however, these levels are so close that they practically coincide. This is because the binding energy of an inner electron is of the order of 1 keV, which is so large compared with that involved in the interactions between atoms that the latter have a negligible effect upon the inner shells.

It is not the same with the electrons in the outer shells just below the valence shell since they are weakly bound to the nucleus. The environment then has a significant effect: the band of energy levels becomes wider by several tenths of an electron-volt as the binding energy decreases.

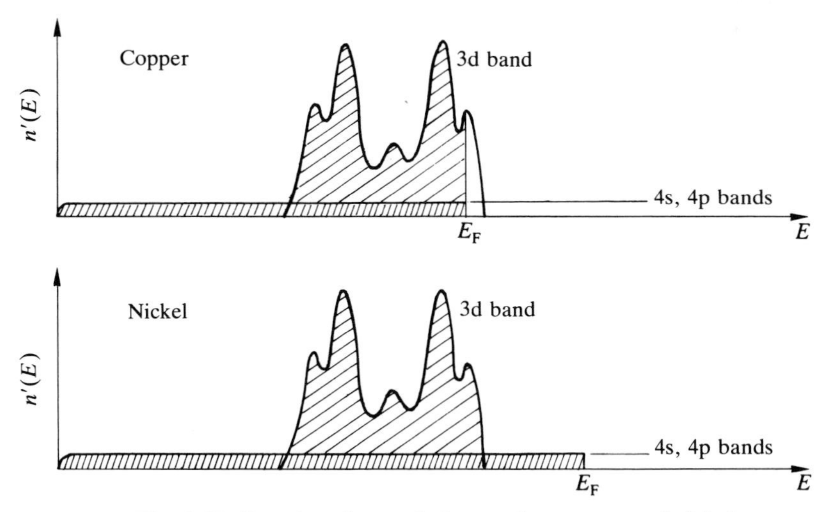

Fig. 2.22. Density of occupied states for copper and nickel.

The band structure is determined by the curve of the density of states plotted against the energy. In general, we can say that the s and p bands (those which arise from the spread of the electronic s and p levels) are very wide and can be described to a good approximation by the free electron theory. The d and f bands, on the other hand, are much narrower and are better described by an approximate calculation from the electronic states of the isolated atom: an attempt is made in this case to evaluate the perturbations caused by the insertion of the atom in a crystal.

Figure 2.22 shows diagrams of the density of states for copper and nickel. Each band is given the same label as that of the corresponding level in the isolated atom. These diagrams show that there is a continuous change from the very narrow level of the electrons in inner shells to the wide band of the valence electron.

Insulators and conductors according to quantum physics

Insulators

The valence electrons of the atoms in a crystal occupy the states in a certain energy band. We deal first with the case in which the following conditions are fulfilled (Fig. 2.23):

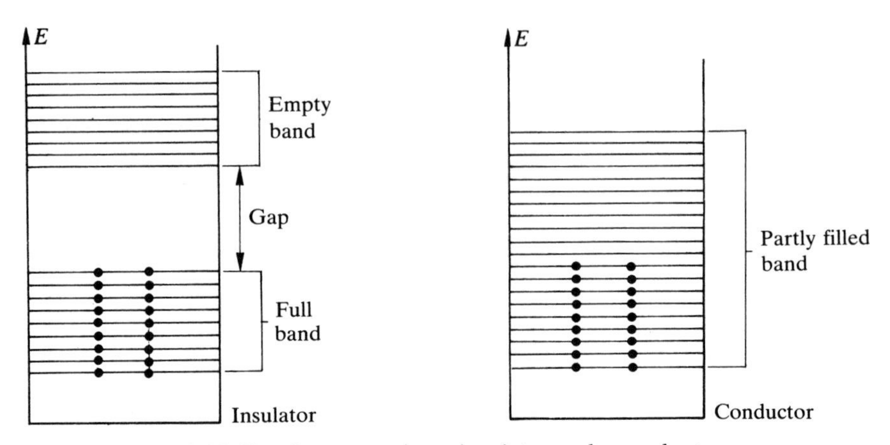

Fig. 2.23. Band structure in an insulator and a conductor.

1. The band is *completely full*: this is what happens when the number of valence electrons is twice the number of levels in the band, i.e. two per unit cell.

2. Above the full band, there is a *forbidden band of considerable width*, i.e. of the order of several electron volts. The allowed band just above the forbidden band is completely empty.

In the quantum picture, the wave associated with the electron is propagated throughout the whole crystal and, as we have seen, the electrons themselves have a high mean velocity. When the valence band is full, the ends of the state vectors $\boldsymbol{k}$ occupy all the possible sites inside the Brillouin zone. However, the zone is centrosymmetrical, i.e. to any state $\boldsymbol{k}$ there corresponds a state with an opposite wave vector $-\boldsymbol{k}$. The electrons therefore occur in pairs with momenta which are equal in magnitude but opposite in direction. As a result of this, there is no macroscopic electric current in the absence of an applied electric field in spite of the great thermal agitation of all the electrons.

If an electric field is now applied to the crystal, nothing appears to happen: an electron can only change state by jumping out of its level into another band, since the first band is full; to do this it has to acquire an energy greater than the appreciable gap separating the two bands. The applied field cannot supply this extra energy, so that the overall state of the electrons is unchanged by the application of the field. Whether or not there is a field, no macroscopic electric current is produced and the crystal is therefore an insulator.

Thus, according to quantum theory, a crystal is an insulator not because there are no mobile and almost free electrons, but because their individual momenta are distributed in such a way that their resultant cancels out, even when a field is present.

Conductors

Suppose now that the valence band is not completely filled with electrons. This is the situation in monovalent metals such as copper and sodium. The first band (the s band) can accept two electrons per atom, whereas there is only one valence electron. The occupied states, those of lowest energy, are inside the Fermi sphere (p. 84) which is a long way from the Brillouin zone boundaries.

In the Fermi sphere, just as in the completely full zone, the distribution of occupied states is centrosymmetrical, and the conclusion is the same as in the case of insulators: there is no current in the crystal in the absence of an applied electric field.

However, the situation is different when a field is applied. Suppose an electric field $\boldsymbol{E}$ is applied for a short time Δt. Each electron then experiences a force $-e\boldsymbol{E}$ and the impulse $-e\boldsymbol{E}\Delta t$ given to it causes its momentum to change by $\hbar\Delta\boldsymbol{k} = -e\boldsymbol{E}\Delta t$. Since there are vacant sites in the zone, jumps can take place from one state to another: all the occupied states are given a translation $\Delta\boldsymbol{k}$ in $\boldsymbol{k}$-space, so that after the impulse from the field these states are included in a sphere derived from the Fermi sphere by the translation $\Delta\boldsymbol{k}$ (Fig. 2.24). The origin of the vectors in $\boldsymbol{k}$-space is no longer a centre of symmetry and so the momenta of the

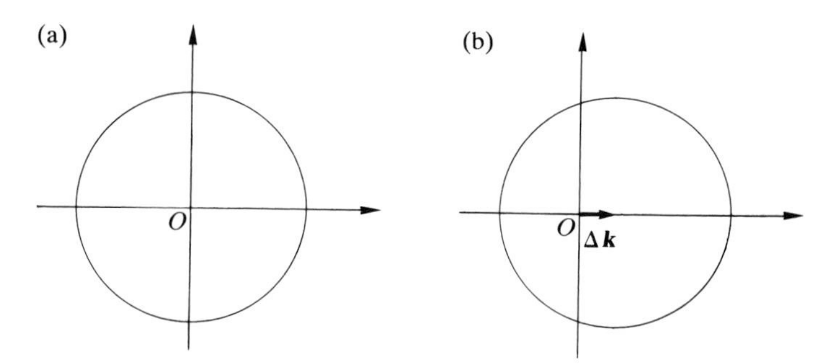

Fig. 2.24. Distribution of electron states in the Fermi sphere: (a) with no applied field; (b) after a short electric field pulse.

electrons no longer exactly cancel out in pairs. The field has thus created an overall drift current of the electrons in the direction of $\Delta\boldsymbol{k}$, i.e. parallel to the applied field $\boldsymbol{E}$: the crystal is a *conductor*. The reason for the conductivity is that the valence electrons do not completely fill the Brillouin zone.

Why are divalent metals conductors?

In this case, the two valence electrons per atom completely fill the first Brillouin zone, so that divalent metals should be insulators: placing two electrons in each atomic level occupies every available state in the band up to those with the maximum energy. However, we have already explained (p. 94) that the bands in metals arise from a spreading out of the atomic levels. The s and p bands are generally very wide indeed and overlap each other over a large energy range. Because of the degeneracy of the atomic states (2 for s states, 6 for p states), these bands can receive up to 8 electrons per atom and are thus only partly filled by the two valence electrons in a divalent metal. We thus arrive at a situation similar to that of the band in monovalent metals.

Crystal imperfections and the electrical resistance of metals

A problem arises when we ask what happens after the impulse received from the electric field has ceased. As we have mentioned several times, the electron wave is freely propagated in a perfect crystal with its atoms at rest: each state, specified by the vector $\boldsymbol{k}$, is a stationary state, so that the distribution of states must stay constant in time, as it has been created by the impulse from the field. Once started, the current will continue without external excitation. In other words, this first attempt at explaining conduction from the quantum mechanical point of view leads to the prediction that a crystalline conductor will have zero resistance. This is not what is observed (except in superconductors, where the persistence of current is due to a special mechanism — see p. 126).

What is observed in metals is that, as soon as the field is removed, the current falls rapidly to zero. This decrease is described as being due to the electrical resistance of the metal. Furthermore, the disappearance of the current is accompanied in ***k***-space by a spontaneous return to the centrosymmetrical configuration of the electronic states in the absence of a field. The electrons lose energy in this process, since the confinement of all the states within the Fermi sphere is the configuration with the minimum energy.

The resistance of real conductors is explained by the action of geometrical imperfections in the crystal lattice, since no real crystal is perfect. The electron wave may be scattered by crystal defects such as lattice vibrations (phonons), lattice strains, etc. As a result of such scattering, an electron can change its state with a transfer of energy from the electrons to the lattice: this is the origin of the **Joule heating effect**, which is related to resistance.

We shall not enter into a theoretical investigation of these complex phenomena but, assuming that there is such a link between crystal imperfection and electric resistance, we shall show how the idea of a connection between them enables certain special features of the resistance of metals to be given a qualitative explanation.

1. Even if a crystal is as pure as possible and even if any local strains are reduced to a minimum, the atoms never occupy the nodes of a geometrically perfect lattice because of their thermal agitation. Elastic vibrational waves pass through the crystal (p. 13) and these have an amplitude that increases as the temperature rises. It is therefore immediately apparent why the resistivity of pure metals is very small near the absolute zero: the crystal is very nearly perfect. A calculation of the effect of the elastic waves on the scattering of electron waves shows that the resistivity of metals is, to a first approximation, proportional to temperature (see Box 12), which is

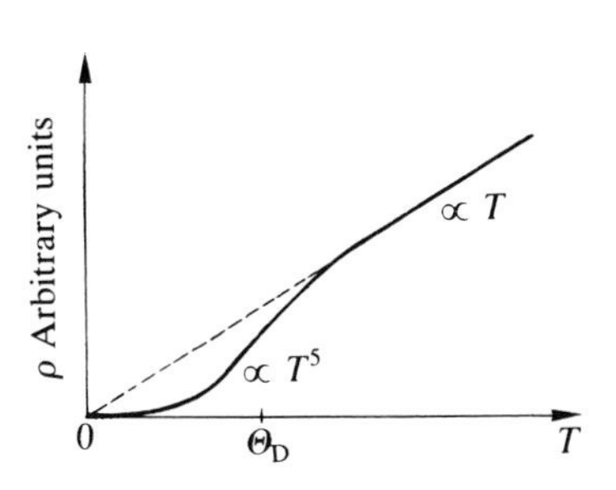

Fig. 2.25. Theoretical curve for the variation of the resistivity of a pure metal with temperature.

in agreement with experiment (Fig. 2.25). This is the result which could not be explained by classical theory.

Box 12

The variation of the resistivity of a pure metal with temperature

When we described the Drude theory, we found an expression for the conductivity of a metal involving the mean time τ between two collisions for an electron:

$$\sigma = ne^2\tau/2m_e$$

where n is the free electron density. This gives a resistivity of

$$\rho = (2m_e/ne^2)/\tau.$$

It is found that this expression remains approximately valid in quantum theory: it is the evaluation of τ which is fundamentally different. Electron motion in a pure metal is hindered mainly by the vibrations of ions: it can be shown that the mean time between two collisions with a set of given objects is inversely proportional to the 'collision cross-section' of these objects.

Because of the thermal agitation, the collision cross-section of ions increases with rising temperature. At temperatures above the Debye temperature Θ_D (p. 24), we know that all the contributions to the energy involved in ionic vibrations become proportional to the temperature. In particular, for the potential energy:

$$\tfrac{1}{2}K\langle x^2\rangle \sim k_B T$$

where K is the restoring force constant (p. 3). This enables us to deduce that the mean square displacement $\langle x^2\rangle$, and thus the collision cross-section of the ion, is proportional to the temperature.

From what we have said above, this means that ρ is proportional to T for temperatures above the Debye temperature. When the temperature tends to zero, departures from linearity are observed, just as the heat capacity deviates from Dulong and Petit's law. A more detailed calculation shows that the resistivity due to ionic vibration varies as T^5 at very low temperatures. The general theoretical curve, valid for any temperature, is given in Fig. 2.25.

2. The replacement of a normal atom by an impurity atom produces a local strain in the crystal lattice which scatters electron waves and thus

gives rise to resistance. At the absolute zero, when thermal agitation no longer occurs, an impure metal has a resistivity, termed its residual resistivity, which depends on the impurity level. For a given type of impurity, the residual resistivity is simply proportional to its concentration.

If a really good conductor is required, the metal used must be of very high purity. The copper or aluminium used in the power lines of the electricity supply network must satisfy this condition and the copper is therefore refined electrolytically.

3. The scattering of the electron waves by impurities and by phonons are two independent effects which are superimposed on each other. The resistivity due to impurities is independent of temperature and is added to that due to phonons which decreases to zero at the absolute zero: this is Matthiessen's rule (Fig. 2.26). A measurement of the ratio of resistivities at two fixed temperatures yields a parameter indicating the purity of the metal. Thus, for standard platinum resistance thermometers, the ratio of resistivities at 100°C and 0°C must be at least 1.39250: impurities would increase both factors of the ratio and would therefore reduce its value.

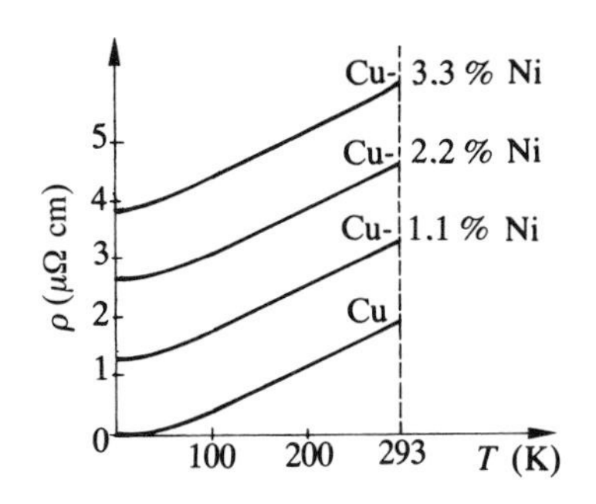

Fig. 2.26. Variation of the resistivity of an impure metal with temperature.

4. For metallic alloys in the form of solid solutions, the lattice nodes are occupied by several different kinds of atom. If the different types are present in comparable proportions, there may be considerable lattice distortion. Alloys therefore have a higher resistivity than pure metals. The wires used in electrical heating equipment must have resistance and for that reason are always made from an alloy (e.g. nichrome, an alloy of nickel and chromium).

The residual resistance of alloys is so large that the variation of resistivity with temperature is relatively small. It can even happen that the resistivity remains constant over a wide range of temperature (e.g. as in constantan, an alloy of copper and manganese).

In solid solutions described as ordered, where the different atomic species alternate regularly throughout the lattice, the crystal is periodic. This is not the case, however, for the same alloy in a disordered state, so the resistivity of the ordered alloy is lower, and this is indeed observed.

5. There are other applications of the association between imperfections and resistivity. A metal deformed by work-hardening (drawing, rolling, etc.) has a higher resistivity than the same metal in a well crystallized state.

A pure metal contains vacancies (SM, p. 107) at high temperatures, which disappear at room temperature. Each vacancy causes a slight increase in resistivity. If a metal is quenched from a high temperature, the vacancies are preserved during the rapid cooling and then gradually disappear as the metal tends towards its equilibrium state at room temperature. This process can be followed by measuring the resistivity of the quenched metal as a function of time. It is true that the resistivity variations are very small, but the measurements can be made with sufficient precision to detect them.

Comparison between classical and quantum models

It is clear that the classical model is simpler and easier to picture as way of explaining the difference between conductors and insulators. In conductors, there are free electrons, which can be set in motion and will flow under the action of an electric field; in insulators, the electrons are bound to the atoms and do not respond to an applied field.

Band theory, which rests entirely on quantum ideas, is much more abstract and less easily appreciated, and that is why the classical model is still sometimes used. In the case of semiconductors, we shall find an example of a dual approach — first classical, then quantum — whereas in describing the electrical properties of, say, sodium chloride, it is obvious that the classical picture of the individual Na^+ and Cl^- ions is preferable.

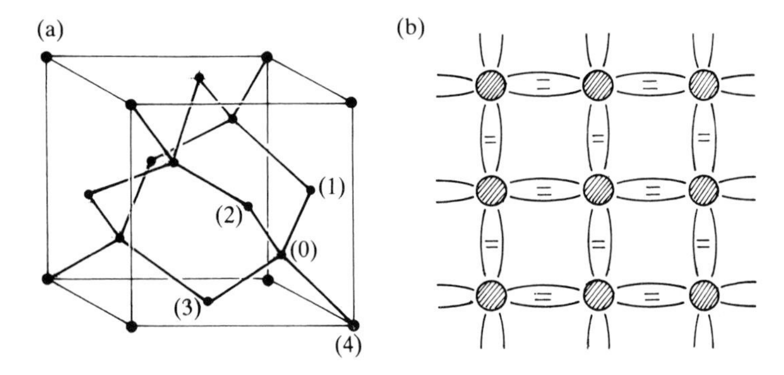

Fig. 2.27. (a) Face-centred cubic unit cell of silicon (2 atoms per cell). Each atom is bound to four neighbours, e.g. (0) to (1), (2), (3), and (4). (b) Diagrammatic representation of the four covalent bonds each with a pair of electrons.

On the other hand, quantum ideas are indispensable if certain simple and basic features of the conductivity of metals are to be explained. This is because the picture of the electron as a material point obeying the laws of classical mechanics is as powerless to describe its behaviour between the atoms of a metal as it is to describe its motion inside an isolated atom.

Nevertheless, the greatest success of quantum physics in connection with the electrical properties of solids lies in the field of semiconductors, an area of study which it opened up and explored. Not only are quantum concepts indispensable to an understanding of the physics of these materials, but they have also been a direct source of developments in solid state electronics.

Semiconductors

Intrinsic semiconductors

The best known semiconductor is the silicon crystal (with germanium a close second). Its crystal structure is the same as that of diamond (SM, pp. 6–7): each atom is linked by a covalent bond to four neighbouring atoms at the vertices of a regular tetrahedron (Fig. 2.27). (This structure is sometimes drawn as if in a plane with the atoms at the nodes of a square lattice. In such a scheme, the number of bonds is correct, but not their directions.)

Diamond is an excellent insulator: the energy required to remove an electron by breaking a bond with the neighbouring atoms is very high (of the order of 7 eV) and up to the highest accessible temperatures (3000 K) there are practically no free electrons in the classical picture.

In silicon, and still more in germanium, the covalent bond is much less strong. An electron is liberated if it acquires, by virtue of thermal fluctuations, an additional energy W (0.55 eV for silicon and 0.36 eV for germanium). According to Boltzmann's law, the probability that the electron is spontaneously liberated is proportional to $\exp(-W/k_B T)$. The number of electrons liberated per atom still remains very small, but it is no longer negligible at temperatures near the melting point.

According to the expression on p. 70, the conductivity is proportional to the number of free electrons and to their mobility. In metals, the number of free electrons is very large (of the order of one per atom) and is independent of temperature, while the mobility decreases as the temperature rises (Box 12).

The important feature in semiconductors, on the other hand, is the very rapid increase in the number of carriers as the temperature rises, so that the slight fall in their mobility can be neglected. It follows that we expect the conductivity σ of a semiconductor to increase with temperature like $\exp(-W/k_BT)$. Pure silicon can be regarded as an insulator at room temperature and a poor conductor at very high temperatures (hence the name semiconductor). According to Boltzmann's formula, a graph of $\ln(\sigma)$ plotted against $1/T$ (Fig. 2.28) should give a straight line whose slope will yield the value of the binding energy W of the valence electrons.†

This is the characteristic behaviour of a pure semiconductor, known as an **intrinsic semiconductor**. The differences from a metallic conductor are quite significant:

1. the orders of magnitude of the conductivities differ enormously: $10^7\,\Omega^{-1}\text{m}^{-1}$ for copper at room temperature and $0.1\,\Omega^{-1}\text{m}^{-1}$ for silicon at 1000 K;

2. the conductivity of a semiconductor increases considerably with rising temperature: for silicon, the ratio of the conductivities at 1500 K and 750 K is 100. The reason for this is the increase in the number of carriers;

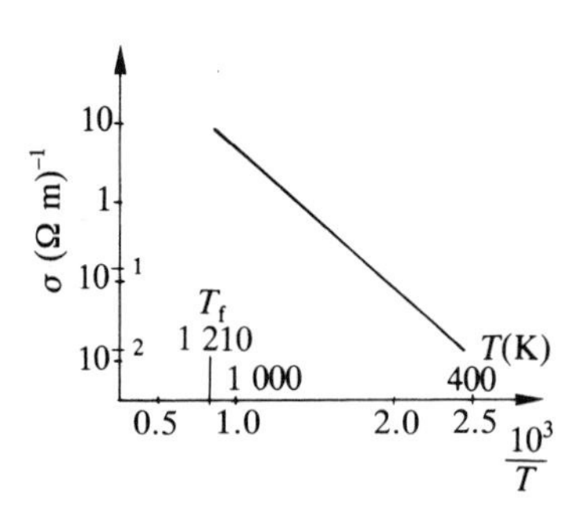

Fig. 2.28. Variation of the conductivity of pure germanium with temperature: $\ln\sigma \propto 0.36/k_BT$. Below 400 K, the intrinsic conductivity is masked by that due to residual impurities.

3. the conductivity of a metal, on the other hand, slowly decreases as the temperature rises: thus, for tungsten between the same temperatures of 1500 K and 750 K, the conductivity is only halved, and this is due to the reduction in the mobility of the free electrons.

Induced conductivity in a semiconductor

Silicon also becomes a conductor if events other than a rise in temperature liberate electrons by breaking their valence bonds. Thus, an electron can become detached by the absorption of a photon or the collision of a high energy particle.

†The parameter W in Arrhenius's law (Fig. 2.28), interpreted classically as the binding energy of the electron, is equal to half the energy gap (p. 90), while the photoelectric threshold energy is equal to the gap.

The maximum energy which can be transferred to an electron when the crystal is irradiated with radiation of frequency ν is that of the photon $h\nu$. For the radiation to be effective, the value of $h\nu$ must be greater than a certain threshold, e.g. 1.1 eV for silicon: the resultant **photoconductivity** is only produced with light of wavelength less than 1.12 μm.

In this case, semiconductors cannot be clearly distinguished from insulators. Even if the energy of the covalent bonds is large, energetic photons (i.e. very short wavelength radiation) can produce photoconductivity. Diamond, for example, in which $(h\nu)_{\min}$ is 7 eV, is sensitive to the far ultraviolet.

The induced conductivity is proportional to the number of electrons liberated, and thus to the illumination of the solid under the given conditions. This property is used in photoresistors and the construction of simple photometers.

Doped semiconductors

We have used the expression 'pure' silicon, but the degree of purity should be more precisely defined. It is much greater than that found in practice in chemicals and metals, where 'high purity' means that the impurity concentration is of the order of 1 p.p.m. (part per million). For semiconductors, concentrations a hundred times less than this have to be achieved. The development of a semiconductor technology has only been possible through the discovery of an adequate purification technique (zone refining). No method of chemical analysis is capable of determining such low impurity levels without great difficulty, so that physical methods, particularly resistivity measurements, have become indispensable.

Once strictly pure silicon was obtainable, it was possible to prepare a whole family of so-called **doped** silicons by introducing additional elements in very small and very precisely controlled concentrations (about 10^{-8} to 10^{-6}), the added atoms replacing those of silicon at the lattice nodes. Doped semiconductors display a great diversity in their electrical properties and it is this which has been the mainspring of the enormous developments in solid state electronics.

n-type semiconductors

Consider first a dopant of valency 5, such as As, P, etc. At a lattice node occupied by the dopant As, the ion has a charge $5e$ instead of the $4e$ for silicon, and it is surrounded by 5 electrons: four are used in the normal covalent bonds made with the neighbouring silicon ions. The site occupied by the arsenic is different from a normal site in that there is an extra charge e associated with it and an additional electron moving around the centre of attraction: a hydrogen-like atom immersed in the silicon is thus formed.

This can be taken into account simply and macroscopically by assuming that the electron is moving in a region whose relative permittivity is that of silicon, which is very high at about 12. It follows that the Coulomb force of attraction experienced by the electron is 12 times weaker; the distances of the electron from the centre of attraction will be 12 times greater than in the free hydrogen atom; and finally the binding energy of the electron to the impurity centre is 12^2 times less than the binding energy of the electron in a hydrogen atom and is of the order of

$$W_1 = 13.6/12^2 = 0.1\,\text{eV}.$$

The additional electrons are thus so weakly bound that, even at room temperature, some of them are liberated. Silicon doped with arsenic possesses almost free electrons 'donated' by the arsenic, which is thus called a **donor atom**. This gives rise to a conductivity which varies with temperature according to $\exp(-W_1/KT)$, where W_1 is the binding energy of the electron to the impurity atom. This conductivity, described as **extrinsic**, is very much greater than the intrinsic conductivity of silicon at room temperature, although it remains very small compared with that of metals. It should be remembered that the maximum number of carriers is equal to the number of arsenic atoms, i.e. 10^{-8} times the total number of atoms. The conductivity produced by donor atoms is known as **n-type** conductivity because the carriers are electrons and thus negative.

The effect of impurities is completely different in metals and semiconductors. In metals, impurities *reduce* the conductivity because they create strains in the lattice which reduce the mobility of free electrons. In semiconductors, the conductivity is *increased* because the donor atoms provide free electrons which do not exist in the strictly pure semiconductor.

p-type semiconductors

What happens if the silicon is doped with atoms having a valency of 3, such as aluminium? To produce the normal bonds between neighbours, it lacks one electron, a defect which is electrically compensated since the Al nucleus has one positive charge e less than silicon. It is observed that silicon doped with aluminium also acquires an extrinsic conductivity, but of a type different from that produced by arsenic: the semiconductor is said to be **p-type**.

A classical model of the kind we have used for n-type semiconductors is not suitable for p-type. It is necessary to call on quantum physics: only then is the symmetry existing between the two types of semiconductor revealed, enabling us to achieve a deeper understanding of both phenomena.

Extrinsic (or impurity) semiconductors and band theory

General solid state theory tells us that the external electrons in the atoms of a solid have states specified by the wave vector $\boldsymbol{k}$ distributed over energy bands. In pure silicon, the electronic states are divided into two bands: the **valence band** and the **conduction band**, separated by an energy gap. At 0 K, the four valence electrons in each atom occupy the whole valence band, while the conduction band is empty. The feature of semiconductors that distinguishes them from insulators is that the gap between the valence and conduction bands is small, of the order of 1 eV or less.

Because the gap is so small, electrons in the valence band can jump into the conduction band at higher temperatures and this obviously involves more electrons as the temperature is raised. Since the valence band is no longer completely full and the conduction band now contains a few free electrons, a drift current is able to flow under the action of an applied electric field: the silicon becomes weakly conducting at high temperatures and this is what we called intrinsic conductivity. We thus recover the same results as those obtained previously using the classical model.

Now take the case of n-doped silicon. The four electrons per atom filling the valence band are supplemented by those provided by the donor atoms. If they were completely free, they would occupy the lowest levels of the conduction band. In fact, they are attracted by the positive centres formed by the donor atoms and this creates new states (Fig. 2.29) located just below the conduction band. At 0 K, the conduction band is empty and the conductivity is zero. But even at temperatures only a little above zero, the additional electrons acquire sufficient energy to jump into vacant places in the conduction band, where they behave like free electrons. These are the carriers responsible for n-type extrinsic conductivity.

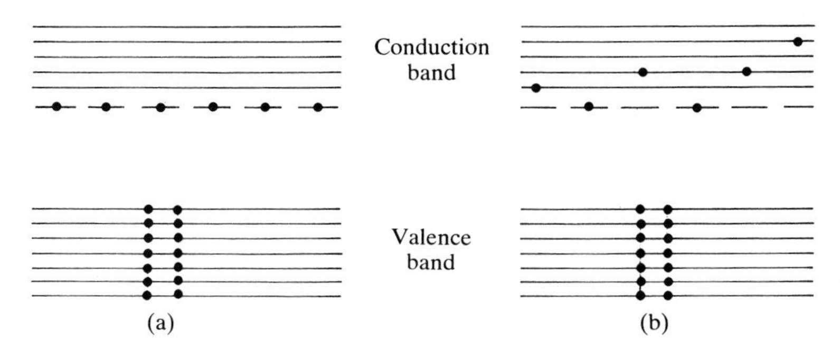

Fig. 2.29. Distribution of electrons in n-type silicon (a) at 0 K: the additional electrons are all occupying levels created by donor atoms; (b) at $T \neq 0$ K.

The conditions governing the motion of these electrons at the bottom of the conduction band are not the same as those for free electrons in the conduction band of a metal. In the latter case, the gain in energy between the state $k = 0$ and a state k is purely kinetic in origin:

$$E = \tfrac{1}{2}m_e v^2 = (\hbar k)^2/2m_e.$$

On the other hand, for states close to the boundary of the Brillouin zone, the difference in energy between the state k and the state of minimum energy k_0 at the edge of the zone is not due only to a gain in kinetic energy. However, just as the energy of the free electron is proportional to k^2, so the energy ΔE at the bottom of the band is, to the first order, proportional to $(k - k_0)^2$ (Fig. 2.30). It can therefore be written in the form:

$$\Delta E = \frac{1}{2m_c}\,[\hbar(k - k_0)]^2.$$

This formula suggests that we adopt the following approximation: the electron behaves as if it were free but with an **effective mass** m_c different from the normal mass m_e.

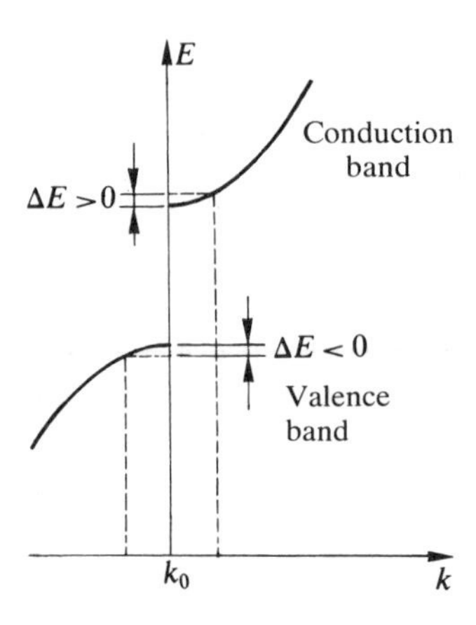

Fig. 2.30. The variation in the energy of states with the magnitude of the wave vector k near the gap between the valence and conduction bands: the variation is a quadratic function of $k - k_0$.

We now move on to the case of silicon doped with an element of valency 3 whose atoms are called **acceptor** atoms. Around each of these there is a deficiency of one electron in comparison with pure silicon: the valence band of doped silicon is therefore no longer completely full. Thus, under the influence of an electric field, the electrons can change state and an electron current can flow. However, the mechanism of this type of conductivity, described as p-type, is distinct from that of n-type conductivity which is due to the presence of a few electrons in the almost empty conduction band.

Consider an electronic state of wave vector $\boldsymbol{k}$ near the edge of the valence band, and with an energy less than that associated with the band edge (Fig. 2.30). The energy difference ΔE is negative and proportional to $(k - k_0)^2$, so that it can again be written in the form:

$$\Delta E = \frac{1}{2m_v}[\hbar(k - k_0)]^2.$$

provided that the effective mass is given a negative sign. When an electron is in a state near the edge of the band, it does not behave at all like a free electron. Because of the stong interactions between the electron wave and the lattice, a force applied to the electron causes it to accelerate in the opposite direction: this is the physical meaning of the introduction of a negative mass.

In the quantum mechanical equations, as in the equations of motion of a particle in classical mechanics, there is a single parameter involved: the ratio of charge to mass or the specific charge. A negative charge, $-e$, with a negative effective mass m_v is equivalent to a fictitious particle with a positive charge e and a positive effective mass $|m_v|$. This is what is called a **positive hole**. The movement of an electron in the top of the almost full band is equivalent to the movement of a positive hole at the bottom of a band 'empty of holes'.

The idea of a positive hole is a product of quantum mechanics and band theory, and that is why classical physics cannot provide a satisfactory explanation of p-type conductivity. The existence of two different mechanisms in n-type and p-type conductivity is an essential feature of the physics and technology of semiconductors. It is to deal with this sort of phenomenon that we wish to go beyond the stage of ideas accepted because they are familiar (or of formulae used because they work) to a point where we emphasize their physical meaning, but without necessarily appealing to the calculations that justify them.

Consider a valence band which would be completely filled by N electrons and which lacks a single electron in an energy level near the band edge. We shall make use of the two following observations:

1. The full band cannot give rise to a current: since its contribution to any current is the sum of that from the band with $N-1$ electrons and that from the isolated electron, these two contributions must be equal in magnitude and opposite in sign.

2. If we unite the electron and the hole corresponding to it, we obtain an object with no charge whose contribution to the flow of current is zero. Thus, the contribution of the electron is opposite to that of the hole.[†]

[†]The hole, in the sense given to it in semiconductor theory, is a particle and not a region of empty space. It is the combination of the electron and the hole which is 'nothing'.

By bringing these two propositions together, we deduce from them that the contribution from the band with $N-1$ electrons is the same as that from the isolated hole. More generally, the band with $N-q$ electrons, provided that q is small compared with N, behaves like a set of q independent holes.

That is how the symmetry between the two types of conductivity emerges. The donor atoms introduce electrons into the conduction band which behave as almost free carriers of negative charge with an effective mass generally different from that of the free electron. The acceptor atoms, on the other hand, introduce holes into the valence band which behave as almost free carriers of positive charge with their own effective mass.

We can go even further with the analogy. When an acceptor atom is introduced into pure silicon, a negative charge (the difference between the charge on the Si^{+4} and Al^{+3} ions) and a positive hole is placed in a lattice site. At low temperatures, the hole is bound to the acceptor atom just as the electron is bound to the donor atom in the n-type material (Fig. 2.31). At 0 K, the holes are located in the levels created by the acceptor atoms. However, the bond energy is small and, at room temperature, the hole is liberated by thermal agitation and delocalized in the lattice. It becomes a positive charge carrier capable of generating a current, just as the electron is a negative charge carrier in n-type semiconductors. The positive hole, however, does not have a familiar classical image as the electron does: hence the difficulty of understanding p-type conductivity.

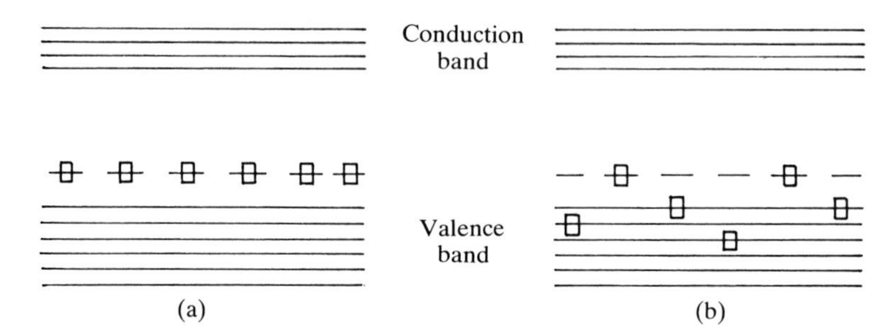

Fig. 2.31. Distribution of holes in p-type silicon (a) at 0 K, (b) at $T \neq 0$ K. Note: the energy changes in n-type and p-type silicon are in opposite directions (cf. Fig. 2.29) because the charges associated with electrons and holes have opposite signs.

Distribution of electrons over the possible states of a semiconductor

We have several times referred in previous sections to the effect of temperature on the distribution of states occupied by electrons in silicon,

whether pure or doped. This statement can be made more precise by using Fermi–Dirac quantum statistics (Box 10).

At 0 K, the situation is simple: the levels arranged in increasing order of energy are occupied at the rate of two electrons with opposite spins per level, until the N electrons of the system are accommodated. The maximum energy level reached is the Fermi level: the probability of occupation of a state of energy E is 1 if $E < E_F$ and 0 if $E > E_F$. There is a sharp drop from 1 to 0 at the Fermi energy.

Above 0 K, the probability of occupation is still 1 for very small values of E and 0 for very large values. The Fermi–Dirac function describes the transition between 1 and 0 around the Fermi energy, and it has the property that the transition zone has a width of the order of $k_B T$, which therefore increases with rise in temperature (Fig. 2.16).

In order to find the distribution of electrons in a given system, it is first necessary to determine the curve giving the **density of possible states** as a function of the energy of the state i.e. to find $n(E)$ such that $n(E)\Delta E$ is the number of states with energies lying between E and $E + \Delta E$. This density of states is then multiplied by the probability of occupation of a state with energy E at the temperature T, this being given by the Fermi–Dirac function. That was what we did for the electrons in a metal (see Fig. 2.15, p. 85). The problem is more difficult to solve in the case of semiconductors (Box 13), for two reasons. Firstly, the density of states is complex because there are the two bands, valence and conduction, separated by an energy gap. Secondly, for the calculation of the electron distribution, the Fermi–Dirac function depends on the parameter E_F which is not known a priori

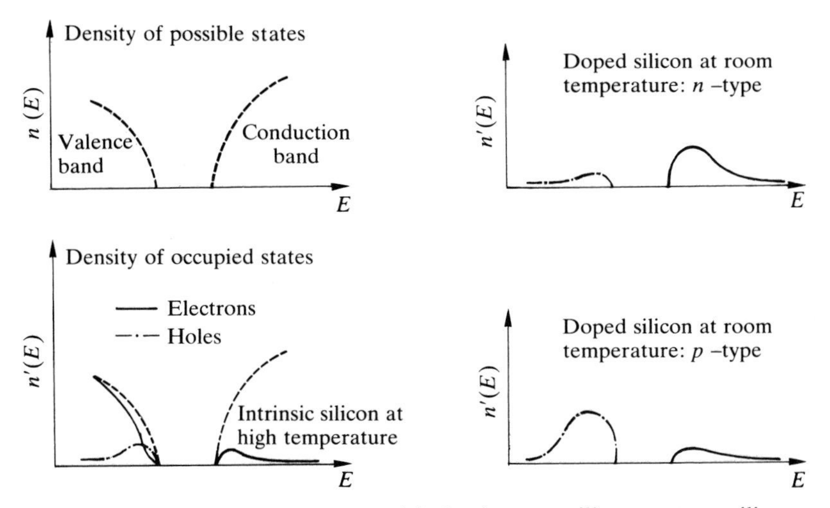

Fig. 2.32. Distribution of electrons and holes in pure silicon, n-type silicon and p-type silicon.

but is determined by the condition that the number of occupied states is equal to the total number of electrons.

Figure 2.32 shows the distribution of electrons and holes in three cases: pure silicon at high temperatures, n-type silicon and p-type silicon. In all cases, there are two types of carrier: electrons in the conduction band and holes in the valence band. The number of holes is equal to the deficiency of electrons compared with the full band. In intrinsic silicon, the number of electrons is equal to the number of holes: in doped silicon, there is a majority carrier, electrons in n-type and holes in p-type.

Box 13

Calculation of the number of carriers in a semiconductor

We first calculate the number of conduction electrons. In an energy range ΔE, the number of electrons is $\Delta n = n'(E)\Delta E$, where $n'(E)$ is the density $n(E)$ of states in the conduction band multiplied by $f(E)$ the probability of occupation of a state with energy E at temperature T (f is the Fermi–Dirac function, p. 85). Thus:

$$\Delta N = \frac{n(E)}{1 + \exp \dfrac{E - E_F}{k_B T}} \Delta E.$$

To find the total number of electrons, it is then sufficient to integrate this over the whole conduction band from the bottom E_c to the top E'_c:

$$N = \int_{E_c}^{E'_c} \frac{n(E)}{1 + \exp \dfrac{E - E_F}{k_B T}} \mathrm{d}E.$$

The difficulty in this calculation arises from the fact that $n(E)$ is poorly known and depends on the nature of the semiconductor. However, in view of the rapidly decreasing form of the Fermi–Dirac function, no great error is introduced if the exact expression for $n(E)$ is replaced by its expansion near the bottom of the conduction band. This expansion generally takes the form:

$$n(E) = \frac{V}{2\pi^2}\left(\frac{2m_c}{\hbar^2}\right)^{\frac{3}{2}} \sqrt{E - E_c}.$$

Except for a different zero of energy (the energy zero being E_c in this case), we recognize here the formula established in Box 8 for free electrons, with the mass m_e replaced by an effective mass m_c. This arises from the fact that, near the edge of the band, the energy varies as the square of k:

$$E = E_c + \frac{\hbar^2 (k - k_0)^2}{2m_c}$$

just as for free electrons.

As a result, we write:

$$N = \int_{E_c}^{+\infty} \frac{V}{2\pi^2} \left(\frac{2m_c}{\hbar^2} \right)^{\frac{3}{2}} \frac{\sqrt{E - E_c}}{1 + \exp \dfrac{E - E_F}{k_B T}} \, dE$$

in which the replacement of upper limit of integration by $+\infty$ has only a minimal effect on the value of the integral since, when $E \gg E_c$, the product $n(E)f(E)$ becomes negligible (Fig. 2.16). To continue the calculation, we make another approximation:

$$E_c - E_F \gg k_B T.$$

This approximation is valid for most normal intrinsic semiconductors at room temperature. In that case, we can replace $f(E)$ by $\exp[-(E - E_F)/k_B T]$, since the 1 in the denominator becomes negligible. The expression for N then reduces to:

$$N = \frac{V}{2\pi^2} \left(\frac{2m_c}{\hbar^2} \right)^{\frac{3}{2}} \int_{E_c}^{+\infty} \sqrt{E - E_c} \exp\left(- \frac{E - E_F}{k_B T} \right) dE.$$

To calculate the integral, we put

$$E = E_c + k_B T x^2$$

and obtain:

$$N = \frac{V}{2\pi^2} \left(\frac{2m_c k_B T}{\hbar^2} \right)^{\frac{3}{2}} \exp\left(- \frac{E_c - E_F}{k_B T} \right) \int_0^{+\infty} x^2 e^{-x^2} dx.$$

Knowing that:

$$\int_0^{+\infty} x^2 e^{-x^2} dx = \frac{1}{2}\sqrt{\pi},$$

we find for the number $n = N/V$ of electrons per unit volume:

$$n = 2\left(\frac{m_c k_B T}{2\pi\hbar^2}\right)^{\frac{3}{2}} \exp\left(-\frac{E_c - E_F}{k_B T}\right).$$

The Fermi level appears in this expression, but because we have not included the overall balance in the number of carriers (number of electrons equal to the number of holes in an intrinsic semiconductor without impurities) the Fermi level is not yet determined. We must therefore calculate the total number P of holes in the valence band, the calculation being completely symmetrical with that of the number of electrons in the conduction band:

$$P = \int_{E'_v}^{E_v} f_v(E) n(E) \mathrm{d}E$$

where
$$f_v(E) = 1 - f(E) = \frac{1}{1 + \exp\left(-\frac{E - E_F}{k_B T}\right)}$$

is the probability of finding a hole (or of not finding an electron) with energy E. Similarly, we shall use the expression for the density of states near the top of the valence band:

$$n(E) = \frac{V}{2\pi^2}\left(\frac{2m_v}{\hbar^2}\right)^{\frac{3}{2}} \sqrt{E_v - E}$$

where m_v is the effective mass of the positive carriers. By noting that the product $f_v(E)n(E)$ quickly becomes negligible when going further down the valence band, we can calculate P as an integral from $-\infty$ to E_v:

$$P = \int_{-\infty}^{E_v} \frac{V}{2\pi^2}\left(\frac{2m_v}{\hbar^2}\right)^{\frac{3}{2}} \frac{\sqrt{E_v - E}}{1 - \exp\left(-\frac{E - E_F}{k_B T}\right)} \mathrm{d}E.$$

We make the approximation $E_F - E_v \gg k_B T$ and we then put:

$$E = E_v - k_B T x^2$$

which finally gives the expression for the number of holes per unit volume $p = P/V$:

$$p = 2\int\left(\frac{m_v k_B T}{2\pi\hbar^2}\right)^{\frac{3}{2}} \exp\left(-\frac{E_F - E_v}{k_B T}\right).$$

We can now determine E_F: putting $n = p$, we find:

$$E_F = \frac{E_c + E_v}{2} + \frac{3}{4} k_B T \ln \frac{m_v}{m_c}$$

This expression confirms that, in general, the position of the Fermi level depends on temperature. However, in the completely symmetrical case in which the effective masses are equal, it lies at an equal distance from the bottom of the conduction band and the top of the valence band. From now on, we shall assume that this is the case ($m_v = m_c = m$). By substituting the expressing for E_F in the formulae for n and p, we then find:

$$n = p = 2\left(\frac{mk_B T}{2\pi\hbar^2}\right)^{\frac{3}{2}} \exp\left(-\frac{G}{2k_B T}\right)$$

an expression in which the energy gap $G = E_c - E_v$ occurs. This quantitative result illustrates the sharp variations in n and p with temperature arising from the exponential factor.

The classical model (p. 101) also predicts that the number of free electrons is proportional to an exponential, $\exp(-W/k_B T)$. The calculation just carried out shows that the classical activation energy is not equal to the magnitude of the energy gap but to half of it.

In the expression for the total conductivity:

$$\sigma = ne\mu_n + pe\mu_p,$$

it is the variations in n and p which will cause variations in σ. By plotting ln (σ) as a function of $1/T$, we should obtain a straight line whose slope will enable the energy gap to be determined (Fig. 2.28).

The formula for n (or p) enables an order of magnitude to be obtained for the carrier concentration in an intrinsic semiconductor. If we assume that G is about 1 eV, and if we take the effective mass to be the same as that of the electron, we find at $T = 300\,\text{K}$ that

$$n = p \approx 4 \times 10^{16}\,\text{m}^{-3}.$$

In fact, $n = p \approx 6 \times 10^{16}\,\text{m}^{-3}$ for silicon and $2.5 \times 10^{17}\,\text{m}^{-3}$ for germanium, which has a smaller energy gap. It can be seen that such a carrier density is much lower than that in 'good' conductors (which is of the order of $10^{27}\,\text{m}^{-3}$). Another essential difference from good conductors is the large variation of n and p with temperature. If the calculation had been carried out at 30 K instead of 300 K, we should have found that $n = p \approx 4 \times 10^{-59}\,\text{m}^{-3}$ instead of $4 \times 10^{16}\,\text{m}^{-3}$!

It will be noticed that the product np is independent of the position of the Fermi level, but depends only on the magnitude of the energy gap G:

$$np = 4\left(\frac{mk_BT}{2\pi\hbar^2}\right)^3 \exp\left(-\frac{G}{k_BT}\right).$$

This is a very important expression since it remains valid for doped semiconductors. Take the case of an n-type semiconductor: each donor atom provides an additional electron. The number N_d of donors per unit volume is equal to the number of excess electrons, i.e. $n - p$, the difference between the number of electrons in the conduction band and the number of holes in the valence band. It follows that $n = N_d + p$, so that the expression for the product np enables us to write:

$$np = p(N_d + p) = n_i p_i = n_i^2$$

where n_i and p_i are the electron and hole concentrations for the corresponding intrinsic semiconductor.

This expression enables n and p to be calculated. In general, it is difficult to produce doping with a concentration lower than $10^{16}\,m^{-3}$, so that in practice we shall always have, in an n-type semiconductor.

$$n \approx N_d, p = n_i^2/N_d \ll n.$$

We therefore see that the free carrier concentration in the semiconductor is directly determined by controlling the doping level.

The ideas we have introduced are difficult to get to grips with because of their basically non-classical character, so let us now summarize them and add a few extra details. Pure silicon is a conductor only at high temperature and even then a poor one. By introducing impurities into the silicon, it can be made a much better conductor even at room temperature. However, depending on the nature of the impurity, there are two types of conductivity, n-type and p-type. In both cases the real charge carriers are electrons, but in n-type semiconductors they behave like free negative charges $-e$ with a mass different from their real mass, while in p-type semiconductors they behave to a first approximation like free positive charges $+e$, called positive holes. The latter are fictitious charges and it should not be thought that they are kinds of positive electrons entering into the constitution of matter.

The drift of electrons and that of holes in the opposite direction produce macroscopic currents in the same direction. They are therefore added together when silicon, if doped with several elements, contains both electrons and holes in comparable numbers. In pure intrinsic silicon, a rise in temperature liberates electrons: in fact, there is a simultaneous creation of a free electron and a free hole (Box 13) and both contribute to the production of currents which reinforce each other.

A large number of semiconductors with a wide range of properties can be fabricated from pure silicon by simple doping operations. It is particularly important to realize that these properties can be determined from theoretical considerations.

In addition, we have, for simplicity, taken silicon as the typical semiconductor, but there are many others: either elements such as germanium, or compounds like gallium arsenide, indium antimonide, mercury telluride, etc. The principles underlying their properties are the same as for silicon, but the parameters which determine them differ from one to another: these parameters include, for example, the band gap between conduction and valence band, the carrier mobilities on which the resistivity depends, etc.

A semiconductor technology of great diversity has evolved from this group of materials and combinations of them, supported by the strength of the underlying theory. We need only mention one example: that of microcomputers, which would not exist without semiconductors. However, it is not our intention to enter into a discussion of the enormous range of such applications: we merely wish to show, using a small number of examples, how the theoretical principles outlined above enable the properties of a few very common devices to be understood.

The p–n junction

Methods have now been developed for doping a silicon crystal by the introduction of the chosen element into a predetermined zone. Using such techniques, it is possible to produce a thin wafer ($100\,\mu m$ thick) consisting of n-type material over one half of its thickness (formed by doping with arsenic) and p-type material over the other half (formed by doping with aluminium), creating what is known as a **p–n junction**. Note that it is technically feasible to reduce the transition zone between p- and n-domains to an extremely narrow region. Moreover, the continuity is perfect: this would not be the case if separate p-type and n-type slices were simply stuck to each other since, even if they were flat and very clean, they would be separated by an oxide layer.

Both halves of the silicon wafer are conducting, but their electronic states are different. The main experimental observation is that, when an external potential difference of a few volts is applied across the junction with one polarity, a current passes through it; while, when the polarity is reversed, the current is negligible. The junction thus acts like a diode current rectifier, the simplest type of solid state electronic component. We shall explain the operation of the diode in terms of the behaviour of free electrons and holes, the charge carriers whose properties were derived from band theory.

In a homogeneous crystal, whether p-type or n-type, the average electric charge is zero everywhere, the excess or deficiency of electrons being compensated by the differences in the charges of the impurity nuclei.

Imagine a homogeneous n-type region adhering to a homogeneous p-type region. The electrons and holes can diffuse from one region to the other, whereas the heavier and bulkier ions cannot do so. Thus, electrons leave the n-type region and penetrate into the p-type region where they are less numerous, and vice versa for the holes. An electrical double layer is thus formed in the transition zone, positive on the n-type side and negative on the p-type side. This double layer, like the charged plates of a capacitor, produces an electric field directed from the n region into the p region (Fig. 2.33), and this retains the electrons in the n region and pushes the holes into the p region. In this way, there is opposition to any motion of the charge which created the layer and, in the absence of an applied voltage, equilibrium is established. This does not mean that there is no motion of charges across the transition zone, but merely that the currents in opposite directions exactly cancel each other out.

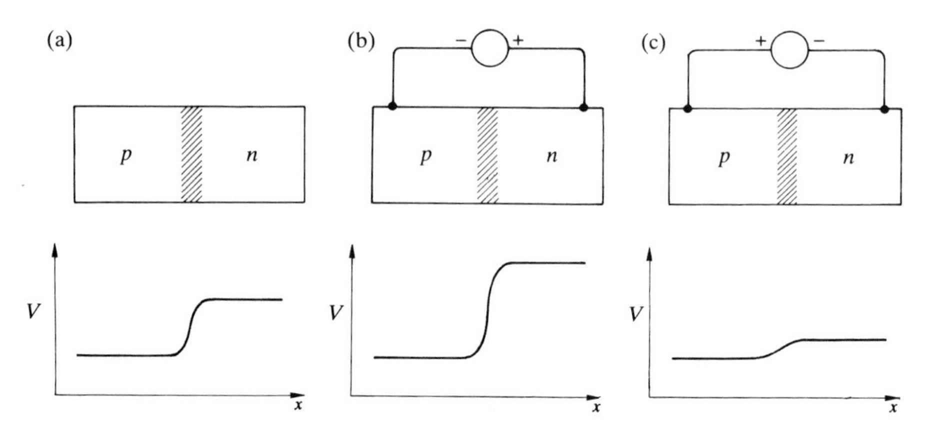

Fig. 2.33. The p–n junction and the variation of internal potential (a) without an external applied voltage; (b) with the n-type region biased positively; (c) with the n-type region biased negatively.

Because of the doping, there are holes in the p region and electrons in the n region. In addition, there are a few electrons in p and a few holes in n (see Fig. 2.32 and Box 13): these are 'minority carriers' in relatively very small concentrations. When the holes in p diffuse into n, they have to surmount the potential difference across the junction and they will do this if they have sufficient thermal excitation. In the other direction, holes in n can easily 'slip' into p at a lower potential, although this current is limited by the small number of holes in n. Such are the two currents which, in the absence of an externally applied voltage across the junction, compensate each other. We give the argument in terms of holes, but it can be transposed to electrons.

It can thus be seen that the inert state of the p–n junction is asymmetrical, so that it is conceivable that an external voltage applied across the junction would have an effect that depended on its polarity. This is what we now demonstrate.

Voltage applied with + polarity over the n region: reverse bias (Fig. 2.33(b))

The external voltage is localized in the transition region, and the potential difference between the p and n regions is therefore increased. Nothing changes for the hole current towards p: the holes are formed in n independently of the applied voltage and 'slip' into p as they did without the applied voltage. On the other hand, the holes in p have to surmount a greater potential difference to penetrate into n: the corresponding current is therefore reduced more and more as the applied voltage increases and it then no longer compensates the reverse current. Consequently, an overall current flows from the + to the − terminal of the external battery. This current tends to a limiting value, that of the hole current towards p. It is very small, because there are very few holes liberated in n by thermal excitation.

Voltage applied with + polarity over the p region: forward bias (Fig. 2.33(c))

The potential barrier between n and p is reduced by an amount equal to the applied voltage V. The probability that a hole will jump from p to n is therefore higher: since it is a thermal excitation effect at temperature T, the current amplification factor is $\exp(eV/k_B T)$. On the other hand, the reverse current has no reason to change and remains as small as ever. In all, there is a current in the direction from p to n, from the + terminal to the − terminal of the battery, which increases very rapidly with the voltage.

The same type of argument can be applied to the electrons: it is found that, for either direction of the voltage, the currents carried by the electrons and holes add together.

To sum up: the junction has a resistance that varies considerably as the applied voltage varies, being conducting when the p region is connected to the + terminal and becoming almost an insulator when the polarity is

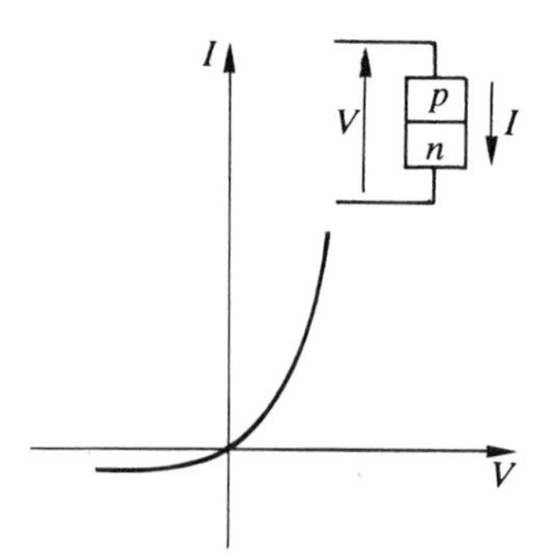

Fig. 2.34. *I–V* characteristic of a p–n diode. The forward-biased current is a few tens of milliamps and the reverse-biased current is of the order of 1 μA for a *V* of several volts.

reversed. The operation of the diode as outlined qualitatively in the above description can be subjected to a detailed calculation: the results of this are in agreement with measurements (Fig. 2.34).

The p–n–p transistor

Localized doping more complex than that of the simple p–n junction can be carried out. We shall describe the best known of these, the **transistor** in the form of a double p–n–p junction. The monocrystalline silicon wafer consists of three layers, of p-type, n-type and p-type material respectively. To give some idea of the orders of magnitude, the thicknesses of the n-type layer may be 5 μm and that of the p-type layers 50 μm. The three layers are called the **emitter** p_1, the **base** n and the **collector** p_2. When there are no

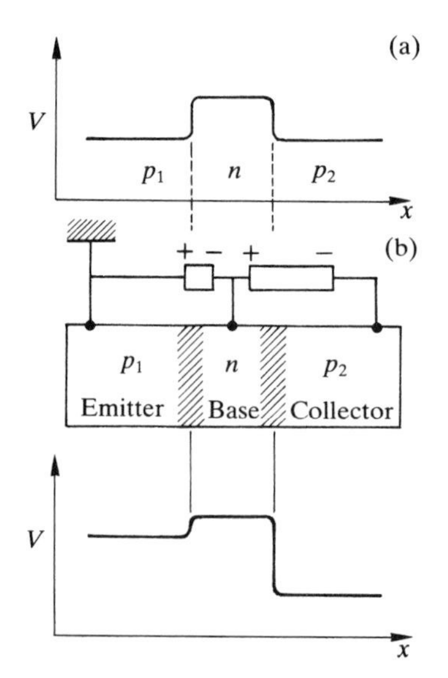

Fig. 2.35. The p–n–p transistor and the variation in potential within it: (a) with no applied voltages; (b) with a positive voltage applied to the emitter and a negative voltage applied to the collector.

external voltages applied between the various regions, the internal potential curve has the general shape shown in Fig. 2.35(a) because the electrical double layers at the junctions produce the effects described in the last section.

If the emitter is biased positively (1.5 V) and the collector negatively (−3 V) with respect to the base, the potential curve of Fig. 2.35(b) is obtained.

The whole system thus appears like two junctions in series: the emitter–base junction is forward biased and the base–collector junction is reverse biased. If the latter were on its own, it would allow only a very small current to pass because of the scarcity of free holes in the n region (see p. 117). On the other hand, the n-type base is supplied with a large hole current coming from the emitter (hence its name) because the first junction is conducting.

The key to the operation of the transistor is to be found in the fact that the thickness of the n layer is small compared with the distance travelled by the holes diffusing in the crystal. Thus, a very high proportion of the holes emitted by the emitter p_1 reach the junction with the collector and pass easily into p_2 since the potential of p_2 is lower than that of n. A very large fraction (more than 95 per cent) of the current from the emitter leaves via the collector, the rest drifting towards the base n. An essential feature as regards transistor applications is that the emitter–collector current I_C is proportional to the emitter–base current I_B. The constant of proportionality, generally denoted β, depends on the transistor geometry and can vary from 50 to 200 for different types. However, when I_B (of the order of 0.1 mA) increases, I_C cannot grow indefinitely since it is in any case less than the ratio of the battery supply voltage to the load resistance in the emitter–collector circuit (Fig. 2.36). The limit is called the **saturation current** of the transistor, and at this limit the potential difference between emitter and collector is very small. The transistor is roughly equivalent to a closed switch. The state can be changed to that of an open switch by cutting

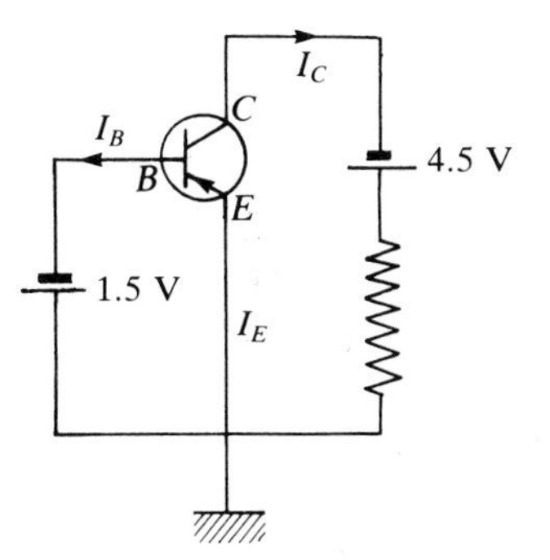

Fig. 2.36. Circuit connections for a p–n–p transistor.

off the base current. Such switching requires only a very small amount of power and can be very fast (a few nanoseconds), thus providing the basis for the extensive applications of transistors in computers.

In the proportional regime $I_C = \beta I_B$, the transistor operates like a current amplifier. It also operates as a power amplifier since the power needed for the control signal is less than the power that can be extracted from the collector current, the extra coming, of course, from the supply battery. High frequency signals (up to 25 MHz) can be amplified in this way.

The field effect transistor: integrated circuits

Consider a wafer of p-type silicon (Fig. 2.37). Two n-type regions, S and D, are produced by diffusion with a suitable dopant. Going from one of these n-type regions to the other, there are two junctions in series but in opposite directions, and hence no current passes from S to D when a voltage is applied. Between S and D, the wafer is covered with a very thin (0.1 μm) insulating layer of silicon dioxide, and this in turn is covered by a metallic layer G. If a positive voltage is applied to G, the minority electrons in the p-type silicon are attracted to the surface and the holes are repelled from it. An n-type channel is thus formed in the surface with, below it, a highly nonconducting region. This n-type channel between the S and D regions allows a current to pass, of a size depending on the structure of the channel and thus depending on the voltage applied to G. G is a 'gate' that controls the amplitude of the current between a 'source' S and a 'drain' D: the electrode G plays the same role as the control grid in a triode vacuum tube. Since the action of G is electrostatic, the power used in the control signal can be extremely low.

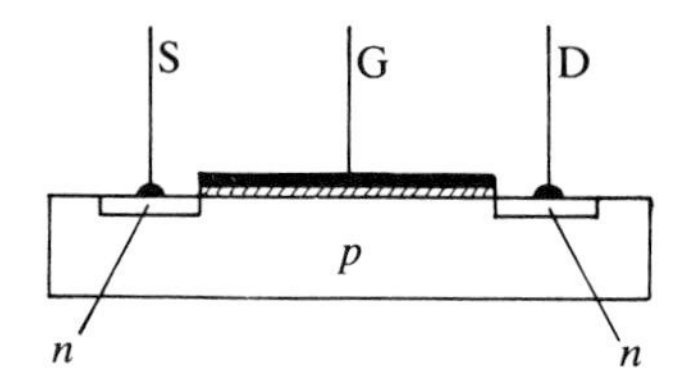

Fig. 2.37. A field effect transistor (a MOSFET).

This type of transmitter is known as MOSFET (for metal–oxide–semiconductor field effect transistor) and all the activity takes place in the very thin surface layer of the wafer. In addition, the dimensions can be reduced to such an extent that the area of the transistor is less than $10^{-3}\,\text{mm}^2$ (miniaturization). A very large number of transistors can be produced alongside each other on a very small wafer (a chip). Such transistors are connected via conductors (layers of metal deposited on the

Fig. 2.38. An Intel 80386 microprocessor showing an area of 1 cm^2 carrying 275 000 transistors.

surface) and it is also possible to interpose either a resistance between two conductors by giving the semiconductor the required resistivity, or a capacitor if the material is almost an insulator. In this way, we arrive at **integrated circuits** consisting of thousands of transistors covering a few square millimetres and having very complex functions (Fig. 2.38). These are the components which have made possible the astonishing achievements of solid state electronics, particularly in the field of computing.

Electronics based on vacuum tubes was developed with great success during the first half of the twentieth century. The second half of the century has witnessed not only the replacement of vacuum tubes by solid state components but an enormous expansion in the potential applications of electronics way beyond the capacity of the older devices.

Let us analyse the reasons for this revolution. The solid state components which have been invented are not only capable of fulfilling the functions of vaccuum tubes but have several other important advantages over them:

1. The supply voltages are only in the region of a few volts instead of being over 100 volts. This is because the solid state devices are very small, of the order of a few micrometres across, so that very low potential differences are sufficient to produce appreciable electric fields. This is unlike vacuum tubes, in which the various components such as the filaments, grids and anodes are separated by several millimetres. They therefore need to be supplied with higher voltages if the fields acting on the electrons are to be strong enough.

2. In a vacuum tube, the electrons are produced by thermionic emission and this entails a wastage of power in heating the filaments to the necessary temperature. Not only that, but the electrons must be emitted into a vacuum, and this requires a bulky vacuum-tight enclosure.

3. Solid state electronics is faster because of the much shorter transit time of the carriers in the miniaturized components.

4. Finally, the weight and volume of the two types of device, for a given level of performance, have very different orders of magnitude. Those who remember radio sets with 'valves' will realize this by comparing their size with that of present-day 'transistor' radios.

The first computer, ENIAC, was built in 1944 and contained 18000 vacuum tubes: it weighed 30 tonnes and occupied a very large room, while the time needed to carry out a multiplication was 3ms. Its memory held 20 words, each of six decimal digits. Today, a portable computer is a thousand times faster and has a much greater memory.

Solid state electronics has one essential characteristic which it is important to emphasize. It was created and developed as a result of theoretical ideas: more precisely, the ideas of quantum physics. Classical physics is insufficient for an understanding of the fundamentals even of the simplest devices.

Of course, experimental research and very subtle and sophisticated production techniques were needed to obtain the first components with worthwhile properties. But these efforts could only succeed, and succeed quickly, because they were guided by theory. The transistor did not originate from a discovery made more or less by happy chance.

To ensure the industrial success of solid state electronics, the products had to be both highly reliable and cheap. This could only be achieved with techniques carried out to very high standards and thus at the cost of considerable investment both in personnel and equipment. The necessary financial outlay was forthcoming because it was very quickly realized that the views of physicists were being confirmed and that this opened up the possibilities of very large markets.

Although the architecture of transistors, and even more of integrated circuits, is highly complex, their manufacture can be automated and the components mass produced. Moreover, the useful life of solid state components, which operate at room temperature, is very much longer than that of electronic vacuum tubes. For several decades, we have been witnessing a very unusual phenomenon: solid state components have become capable of better and better performance and at the same time have become ever cheaper.

There are few examples in modern science of such a close and rapidly exploited linkage between purely abstract ideas and their practical realization which has, in a few years, profoundly modified everybody's way of life through the development of computerized data processing and new possibilities in communication.

The conductivity of amorphous semiconductors[†]

Our discussion up to this point has been confined to crystalline materials. Since the 1970s, however, scientists have given increasing attention to non-crystalline materials. These can be prepared in various ways: by deposition from the vapour phase on a cold substrate; by deposition from a glow discharge; by heavy ion bombardment of a crystal; or by rapid cooling from the melt (quenching). In the last of these, what is obtained is called a 'glass'. There are two types of glass:

1. Those in which the number of nearest neighbours of each atom, as measured by X-ray or neutron diffraction, is not necessarily integral, and where there is no obvious sense in which chemical bonds exist between the atoms. This group includes most of the so-called 'metallic glasses' often produced by an ultrafast quench of the liquid metal;

2. Those in which a coordination number, often characteristic of the crystal, is maintained. An example is vitreous silicon dioxide (SiO_2) in which, as in the crystal, each Si is linked to 4O and each O to 2 Si. A two-dimensional representation of the positions of the atoms in the crystalline and non-crystalline materials is shown in Fig. 2.39. We shall confine our discussion here to materials in the second class.

Vitreous silicon dioxide is transparent far into the ultraviolet and is an excellent insulator. If we continue to use the expressions 'valence band' for the range of energies occupied by the bonding and non-bonding electrons

[†]Although this book is limited in principle to dealing with properties of crystals, amorphous substances are important and we are pleased to be able to introduce the reader to material specially written for this book by Professor Sir Nevill Mott.

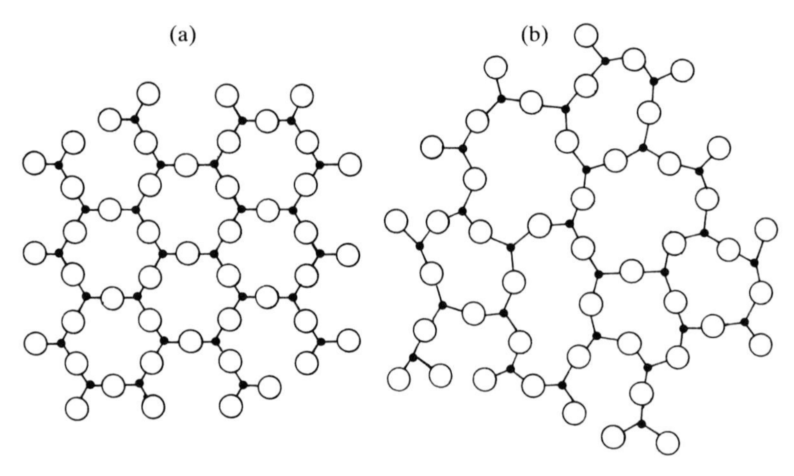

Fig. 2.39. Two-dimensional representation of the SiO_2 structure: (a) in the crystalline state; (b) in the amorphous state.

of the silicon and oxygen atoms, and 'conduction band' for the range of possible energies of an extra electron injected into the material, it is obvious that there must be a wide energy gap between them of the order of 10 eV.

In the band theory of a crystal, we related the existence of the energy gap to that of the Brillouin zone and thus to the Bragg reflection of electron waves (see p. 89). However, in a non-crystalline material, there are no sharp Bragg reflections for electrons or X-rays and the Bragg law does not apply. We must therefore conclude that the existence of a gap *need* not depend on Bragg reflection within the crystal. It may well be that the energy required to remove an electron from one atom, say in the liquid phase of an inert gas, and to relocate it in a distant atom, does not depend on whether the position of the atoms is or is not on a lattice.

So we believe that the concepts of conduction and valence bands can be used for non-crystalline materials. There is, however, a difference. Following theoretical work by Anderson (1958) and Mott (1967), it is believed that, in a non-crystalline material, the lowest states in the conduction bands turn into traps: the electrons are said to be 'localized'. An electron in a localized state cannot move without the help of energy given it by phonons. And since, if we neglect phonons, an electron is therefore 'stuck' or 'not stuck', there must be a sharp energy dividing the two classes of state. This is called a 'mobility edge'. The assumed density of states $n(E)$ in an amorphous material, with a mobility edge E_c, is shown in Fig. 2.40.

A very important non-crystalline semiconductor in which a mobility edge has been identified is hydrogenated amorphous silicon. This material can be obtained in the form of thin films by the deposition of silicon on a

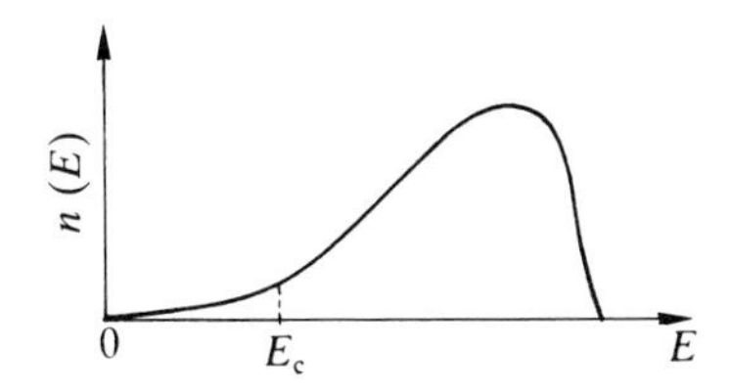

Fig. 2.40. Density of states in the conduction band of a non-crystalline material: E_c is the 'mobility threshold'.

substrate from a glow discharge of silane (SiH_4); it contains several per cent of hydrogen. Most of the silicon atoms have the same fourfold coordination as those of crystalline silicon, which has the diamond structure, but some have a threefold coordination, or less. As a result, there is a 'dangling bond': a silicon bond containing only one electron, giving an electron spin resonance signal. When the material is used as a photoconductor, this provides a very effective electron–hole recombination centre. The hydrogen plays a useful role in passivating most of the dangling bonds by forming Si—H bonds. Without hydrogen, amorphous silicon is a very poor photoconductor.

Experiments have been made to measure the drift velocity and hence the mobility μ of a pulse of electrons excited optically into the conduction band (or symmetrically of holes into the valence band). This turns out to be thermally activated, being of the form:

$$\mu = \mu_0 \exp(-\Delta E/k_B T).$$

ΔE is usually interpreted as the energy interval between the bottom of the band and the mobility edge (although other interpretations have been proposed).

A pioneer in the study of non-crystalline semiconductors is the school of B. T. Kolomiets in the Ioffe Institute in Leningrad. He worked on the so-called chalcogenide glasses formed from As, Se, Te, Si, and Ge. His work showed that, quite unlike crystals, these materials could not be doped: the conductivity was fairly insensitive to composition. The accepted explanation is the so-called $8-N$ rule, where N is the number of outer electrons in a given atom (6 for Te, 5 for As, 4 for Si or Ge). The rule, for which there is some evidence from neutron diffraction, states that, in a glass, each atom is surrounded by $8-\mathrm{N}$ nearest neighbours (2 for Te, 3 for As, 4 for Si or Ge). All the electrons are therefore either bonds, or low-lying *s* states, or in the so-called 'lone-pair' orbitals, which do not take part in bonding. There are thus no loosely bound electrons: this is not possible in a crystal.

In 1975, Spear and Le Comber showed, rather surprisingly, that amorphous silicon *could* be doped, by including PH_3 or B_2H_6 in the silane glow discharge. Amorphous silicon is not a glass, since it cannot be obtained from the melt. It seems likely that the $8-N$ rule is widely valid in true glasses. In amorphous silicon, some of the phosphorus goes in with threefold coordination; two electrons then remain in low-lying *s* states and three in *p* states which form bonds with a neighbouring silicon, so these atoms are not electrically active. But other P atoms go in with fourfold coordination, as in the crystal. The reason why they can do this in spite of the $8-N$ rule appears to be that the fifth electron does not stay with the phosphorus but falls to a dangling bond site, which thus becomes negatively charged. This makes the state more stable. But doping will, therefore, raise the Fermi level from near mid-gap, where it normally resides. Exactly the opposite can be done by adding B_2H_6.

It is therefore possible to construct films of amorphous silicon, with the kind of p–n junction described on p. 115, within the film parallel to the surface. A p–n junction, whether in a crystalline or amorphous material, generates currents under illumination because of the voltage drop to which electrons and holes are subject. The efficiency depends on the lifetime and mobility of the carriers and here crystals have the advantage. But by passivating dangling bonds with hydrogen (and in some cases with fluorine), long lifetimes have been obtained, together with stability and photoelectric efficiencies of the order of 10 per cent. Further work may well improve this value.

At present, amorphous silicon solar cells are used mainly in the familiar pocket calculators and similar devices, but there is real hope that, in the future, they may become important as sources of electrical power on a large scale. Their advantage over crystals is mainly economic; it is possible to cover many square metres with these thin amorphous films relatively cheaply, and mass production could bring down the cost to an economic level. If this should prove to be the case, and large areas of arid land in sunny climates are put to use in this way, storage of electricity may turn out to be the main problem. One possibility may be the generation of hydrogen by the electrolysis of water.

Superconductivity

The resistance of a metal continually decreases as its temperature falls. As the absolute zero is approached, the resistivity tends to a limiting value which becomes smaller as the metal becomes purer and as its lattice is subjected to fewer perturbations (p. 99).

These effects were being studied in detail, particularly at the specialist laboratory in Leyden (Netherlands), at a time when physicists were first able to explore the low temperature region of a few K as a result of the liquefaction of helium. During these investigations, a spectacular and completely unexpected fact was discovered in 1911 by Kamerlingh Onnes (Nobel prizewinner, 1913): the resistivity of mercury fell abruptly to zero and remained *strictly zero* at lower temperatures. Some twenty metals are now known to have a sharp transition to what is called the **superconducting state** at critical temperatures T_c which vary but are always of the order of a few K (7 K for lead, 18 K for niobium, etc).

Some alloys are also superconducting: an enormous number of these have been studied in the search for higher critical temperatures, but the results have been disappointing: it has proved impossible to exceed the record of 23.2 K obtained in 1970 with the compound Nb_3Ge.

Superconducting organic materials have recently (1980) been discovered and these are only poorly conducting at temperatures above T_c. It seems that the two properties of being a good conductor at room temperature and being a superconductor at very low temperatures are by no means correlated. Another example of this is the fact that good conductors like silver and copper are not superconducting even at the lowest temperatures at present attainable (below 10^{-3} K).

The situation was revolutionized in 1986 by the discovery of a complex oxide of lanthanum, barium, and copper which is superconducting up to 35 K (Georg Bednorz and Alex Müller, Nobel Prize winners, 1987). As soon as this was announced, laboratories all over the world began testing oxides with slightly different formulae and the results were dramatic: in 1987, a T_c of 93 K (i.e. above liquid nitrogen temperature) was obtained with the compound YBa_2Cu_3O, and the even higher value of 125 K was eventually achieved. Some research teams announced an observation of superconductivity at room temperature, but the effects were only transitory and not reproducible.

The oxides involved in these latest developments have several features in common. They have a metallic type of conductivity at room temperature; they have similar crystal structures, the perovskite struture, with planes of Cu and O ions incorporating oxygen vacancies and separated by layers containing the other metals. Their conductivity is high in directions parallel to these planes and low in the perpendicular direction. The other common feature is that the copper ions have a 'mixed' valency, i.e. they are a mixture of Cu^{++} and Cu^{+++} ions.

Experimental demonstrations of superconductivity

A superconductor can carry a current with no energy loss at all, although it should be added that this is only true if the current density in the superconductor is less than a critical value J_c which depends on the material (e.g. for NbTi, $J_c = 10^7$ A cm^{-2} and, with one of the new

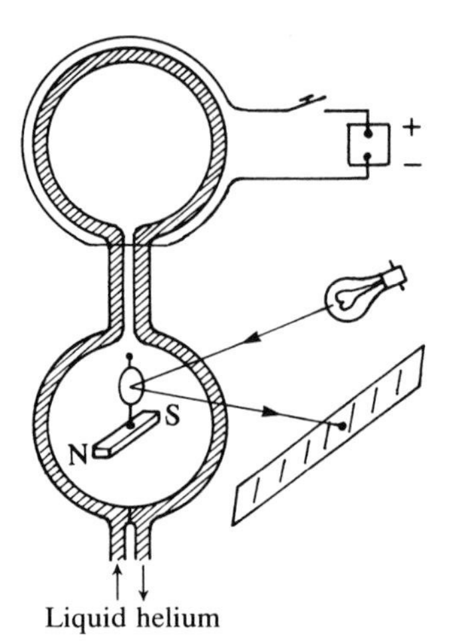

Fig. 2.41. Demonstration showing the existence of superconductivity (Roubeau's apparatus).

superconductors, 10^6 A cm^{-2} has been achieved with the material in the form of a thin film). Potential applications often depend on the value of J_c.

1. Zero resistivity can be demonstrated by the following experiment. Consider a tin-plated copper tube within which liquid helium can circulate (Fig. 2.41). The tin, which becomes superconducting at 3.7 K, forms a closed conducting circuit with two loops: at the centre of the lower loop there is a small suspended magnetized needle, forming a very crude moving magnet galvanometer. Facing the upper loop and adjacent to it is a coil connected to a battery through a switch. Opening or closing the switch induces an emf in the tin circuit.

Above the critical temperature, the galvanometer records a short pulse of current at each change in the magnetic flux linking the circuit (a normal demonstration of electromagnetic induction). The conduction electrons, accelerated by the electromotive induction field, all acquite a drift velocity. As soon as the motive force stops, the electron velocities recover their equilibrium distribution through scattering by the irregularities in the crystal lattice (phonons, strains, etc). The curent thus stops almost instantaneously under the conditions of this experiment.

With the current flowing steadily in the inducing circuit, the tin is continuously cooled until it becomes superconducting at 3.7 K. When it has reached this state, the current in the inducing coil is switched off: the galvanometer is deflected and the deflection becomes steady. The induced current is *permanent* and flows indefinitely in the tin circuit without any external source of power. Even in the most accurate experiments, no evidence has yet been found of any decrease in the current flowing in a superconductor: we must therefore assume that the resistance is strictly zero.

The conduction electrons were excited by the pulse of emf, but the process by which they return to their equilibrium distribution does not operate in the superconducting state. If the temperature of the tin is allowed to rise, this process becomes operative once more as soon as the critical temperature is reached and the induced current stops at once.

2. The superconducting state is characterized not only by the elimination of all finite resistivity, but it is also accompanied by a magnetic effect, the **Meissner effect**. A superconducting metal is not generally a magnetic material in its normal state, so that when it is placed in a magnetic field the field lines pass through it without interruption (Fig. 2.42(a)). However, below the critical temperature, the superconductor becomes magnetized in a direction *opposite* to that of the applied field to such a degree that the field inside the material is reduced to zero: the lines of force are expelled in the superconducting state (Fig. 2.42(a)).

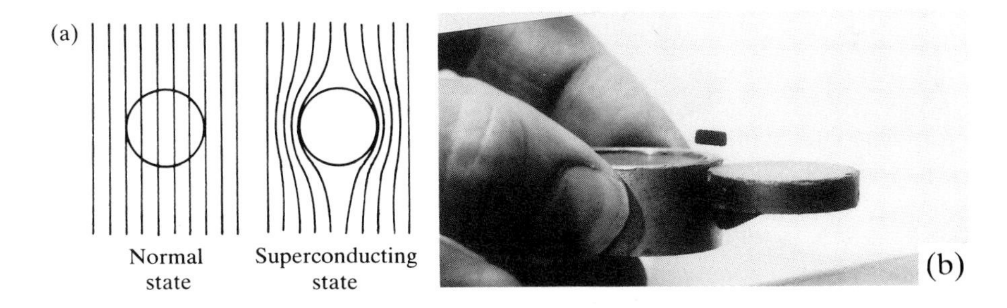

Fig. 2.42. (a) The Meissner effect: magnetic field lines pass through the metal without being affected when it is in the normal state and are expelled from the interior when in the superconducting state. (b) The converse of the floating magnet experiment: a superconducting sample of yttrium barium copper oxide floating above a permanent magnet. The sample has a critical temperature of 95K. (Courtesy of University of Birmingham Superconductivity Research Group.)

This effect is demonstrated by the 'floating magnet' experiment, which has now become relatively simple to perform with the newest materials that are superconducting at liquid nitrogen temperatures. When a small magnet is placed near a horizontal superconducting plate, it is repelled by the magnetization induced in the superconductor as a result of the field from the magnet itself. The repulsive force is in the opposite direction to that of the weight of the magnet and causes it to rise. There is an equilibrium position in which the slightly reduced magnetic force is equal to the weight, so that the magnet floats in space by 'magnetic levitation'. Conversely, if a small superconducting specimen is placed above a strong permanent magnet, it is also repelled by the magnetization induced in it below its critical point. The specimen is then levitated, just as the magnet was in the previous version of the experiment (Fig. 2.42(b)).

There is a limit to the Meissner effect. At any given temperature below T_c, the internal magnetic field is only reduced to zero if the external field is less than a critical value H_c. For fields greater than this, the material returns to the normal state even though its temperature is below T_c. The magnitude of H_c is an important factor in practical applications of superconductors.

The theory of superconductivity

Such a remarkable phenomenon was a challenge to physicists for a long time. They were unable to explain it using classical theory and, from the quantum point of view, the problem is difficult because it is a cooperative effect involving a large number of particles.

The first fruitful approach was phenomenological, i.e. one that made no attempt to find an explanation based on an atomic model. London (1935) showed that a modification to Maxwell's electromagnetic equations inside superconductors could account for the association of the two characteristic properties: the absence of all resistivity and the Meissner effect.

It was not until 1957 that Bardeen, Cooper, and Schrieffer (Nobel prizewinners, 1972) conceived of the microscopic model which contributed what is now considered to be the definitive solution to the problem. However, it is applicable only to superconductors with very low critical temperatures: in spite of a great deal of research, there is not yet a generally accepted theory for the new superconductors with high critical temperatures.

The 'BCS' theory is very complex and is outside the scope of this book. We shall only try to describe a few of the ideas which throw some light on the origin of superconductivity. What we need to understand is how electrons can move through a lattice disturbed by thermal vibrations (small, maybe, but not zero) without losing energy in collisions.

The first feature of the BCS theory is that the electrons form *pairs* known as *Cooper pairs*. There is Coulomb repulsion between two electrons since they have charges of the same sign, so that if they are to pair up there must be an attractive force sufficient to compensate the repulsion. Such an interaction occurs through the intermediary of lattice phonons. An electron passing near positive ions will attract them and disturb their configuration and their vibrations. A second electron, also passing nearby, is sensitive to this deformation and can take advantage of it to reduce its energy. The result is a pair interaction with an energy of the order of a milli-electron volt. An analogy is sometimes useful in understanding the formation of pairs, even though it is undoubtedly far from perfect. Consider a set of particles with identical masses and electric charges of the same sign sliding without friction over the surface of a polished horizontal plate. The particles all repel each other. Now replace the rigid plate with a rubber sheet flexible enough to be slightly depressed by the weight of a particle, so that a small depression is hollowed out around it. Another particle passing nearby might fall into the potential well in spite of the electrical repulsion from the particle causing the depression.

Accepting the formation of Cooper pairs in this way, let us pass on to the second feature of the BCS theory: that the pairs are condensed into the same quantum state and that this minimum energy state is separated from excited states by a gap due to the pair interaction energy. *At very low temperatures*, the energy of thermal vibrations is not large enough to enable this gap to be crossed. As a result, the pair cannot be split up and collisions it suffers cannot change its energy (collisions are perfectly elastic). During the motion of the pair, there can be no transfer of energy to the lattice or, in other words, there is no Joule effect and the resistivity is zero.

On the contrary, for the conduction electrons in a normal metal, those in excited states (p. 95) can pass from one energy level to another that is extremely close, if it is vacant. The electron is then able to transfer its excess energy continually to the lattice: a Joule effect occurs in this case and hence there is electrical resistance.

What emerges from the process described above is that superconductivity can exist only at very low temperatures, and that is why theoreticians discouraged those attempting to find alloys with high critical temperatures. Furthermore, the conditions for the attractive force between the electron pair to be great enough are rarely encountered: there are relatively few superconductors.

And now, suddenly, the new superconductors appear on the scene, confronting theoreticians with a whole series of difficult problems. Is the BCS theory still valid? How can the strong interactions between the electrons in a pair be explained? And so on. For the moment, no convincing and general answers are yet available to such questions.

Predictions of the BCS theory confirmed by experiment

At present, the BCS theory is an indispensable guide in research work on superconductivity and several of its predictions have been confirmed. We mention two simple examples.

At first sight, it appears paradoxical that materials which are very good conductors at room temperature are not superconductors. However, the very fact of their low resistance accounts for this, because the interactions between conduction electrons and the ionic lattice must be particularly weak. This means that the interaction energy of the Cooper pairs will be low and the metal has less chance of being a superconductor.

The second example concerns what is known as the isotope effect. It is observed that the critical temperature of a superconductor varies with the mean atomic mass of the isotopes forming it. Thus, for mercury, T_c changes from 4.85 K to 4.46 K when the mean isotopic mass changes from 199.5 to 203.5; more precisely, it has been found that the critical temperature is inversely proportional to the square root of the isotopic mass. Now it turns out that the same type of variation with ionic mass is exhibited by the Debye temperature (p. 24), so for different isotopes, T_c is proportional to Θ_D. Superconductivity certainly has something to do with the lattice vibrations, which is what is predicted by the BCS theory.

Applications of superconductors

The Joule heating effect in conductors, arising from their resistance, is a cause of energy wastage in all electrical equipment and in carrying electricity over large distances. In spite of the high efficiency of the generating and transmission machinery, the energy wasted over the whole mains network is substantial in absolute terms, so that the possibility of using conductors without losses is of considerable industrial importance. Unfortunately, for superconductors that need to be cooled with liquid helium, the cost of running the necessary cryogenic plant is often greater than the saving that could be made by the elimination of Joule heating. However, the situation has been completely changed by the new superconductors, which only need cooling with the much cheaper liquid nitrogen. The passionate interest aroused by this discovery is the result of the vast perspectives immediately opened up in possible fields of application.

Nevertheless, long and difficult technological research is still needed before it is possible to pass from laboratory experiments to reliable industrial products. Oxide superconductors are ceramics and are thus difficult to work; they are also frequently subject to degradation by their environment (water vapour, carbon dioxide).

The first applications are expected to be in the field of computing (the miniaturization of the components is at present limited by the need to dissipate Joule heat) and in the production of intense magnetic fields (for NMR scanning in medical applications, as in Fig. 2.43, and for magnetic levitation in transport systems). For the time being, however, these remain aspirations, albeit reasonable ones.

The older superconductors such as Nb_3Ge and NbTi, on the other hand, are already used in the laboratory for the production of very strong magnetic fields. When fields much greater than 1 T are needed, iron-cored coils become inefficient and we have no other option than to rely on fields generated by large electric currents (p. 166). In such a situation, with

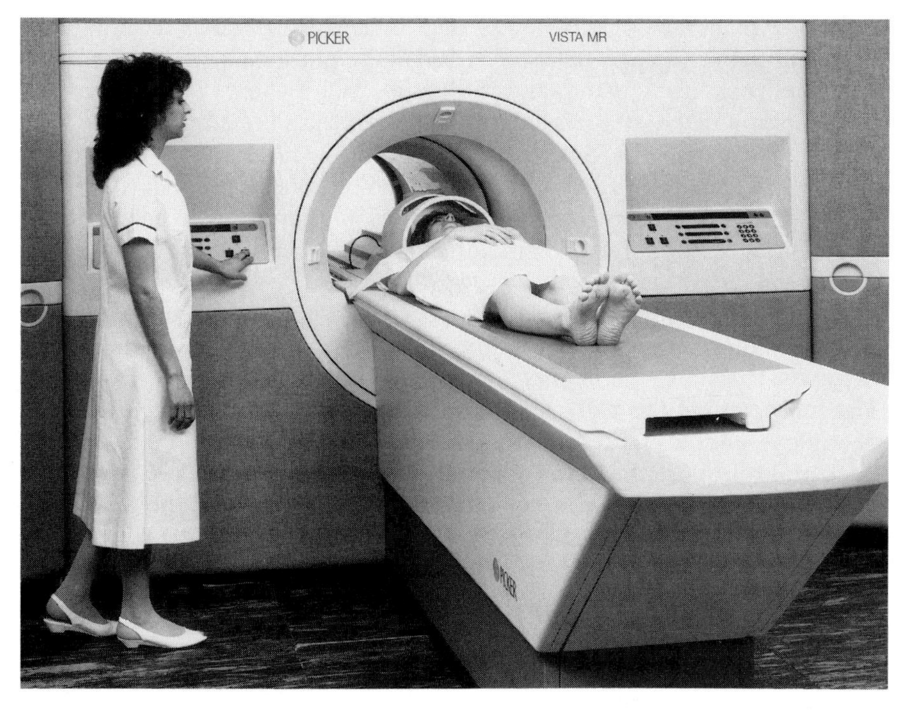

Fig. 2.43. A magnetic resonance body scanner. Machines at present installed in hospitals already use superconducting coils to attain the necessary high field. It has been estimated that changing from helium to nitrogen for the cooling system could reduce costs by £17 000 a year. (Courtesy of Picker International.)

steady fields and currents, all the energy consumed in the windings is dissipated as heat, so that very bulky water-cooled conductors have to be used.

If the windings are superconducting, no energy is needed to maintain the magnetic field indefinitely. The advantages of this are so great that the constraints and the cost involved in installing a liquid helium plant are acceptable, particularly in large particle accelerators. As we have already mentioned, there is a critical field above which a metal ceases to be superconducting, but materials have been found which enable fields of 15 T to be produced.

In a superconducting coil producing a very strong magnetic field, the energy density is considerable, and is of the same order as that contained in a liquid fuel. The metallic components are subjected to enormous mechanical stresses. If, because of an accident, the temperature were to rise and destroy the superconductivity, dangerous explosions could occur in the windings. Safety considerations thus require special precautions to be taken in the construction of large superconducting coils.

The development of our knowledge in the field of superconductors is typical of the way progress has occurred in the physics of solids. The initial discovery was unexpected and for many years there was merely the accumulation of experimental data. It was not until the advent of the BCS theory that systematic advances in the depth of our understanding could begin to take place. In spite of that, it should be pointed out that the theory does not start from 'first principles': it does not permit us to calculate the critical temperature of a given metal. The best superconductors have been discovered by trial and error.

The 1986 discovery has, also unexpectedly, opened up a new area that holds out the promise of a wealth of applications. However, because it currently lacks any theorectical support, it is still at the empirical stage. The full potentialities will only be rationally exploited with the advent of new ideas and, if success is to be achieved, there must be a parallel large-scale effort on the purely technological side.

The whole field provides a good example of the need for the simultaneous development of theoretical and experimental research in the physics of solids.

New conductors

It has long been assumed that good conductors are metals because of their free electrons. Since the 1970s, however, *organic molecular crystals* have been discovered with conductivities approaching those of metals. We shall show with the aid of examples how the structure of these solids explains their unusual properties.

One-dimensional conductors: TTF–TCNQ†

The two molecules, both quite flat, are stacked in columns of TTF and TCNQ alternating regularly in the crystal. In each molecule, there are electrons whose wave functions considerably overlap those of adjacent molecules. On the other hand, there are very few contacts between one chain and another. Thus, the electrons are delocalized along the chain with wave vectors parallel to the axis of the chain: the discrete energy levels form a one-dimensional band, which would just be filled by electrons from all the molecules in the chain.

But one important event occurs: there is charge transfer from the TTF chain to the TCNQ chain. This is an experimental fact: it has even been possible to measure that on average one TTF molecule donates 0.59

†A tetrathiofulvalene–tetracyanoquinodimethane complex

electron to the TCNQ acceptor. It follows that the energy bands are not full: thus, under the influence of a electric field along the axis of the chain, electrons are promoted over the more energetic states. A current is created, while the crystal is an insulator in the perpendicular direction because electrons cannot pass from one chain to another. The conductivity of TTF–TCNQ is $5 \times 10^4 \Omega^{-1} m^{-1}$ along the axis of the chain and 100 times less in a direction perpendicular to the axis.

In reality, the electrical properties of this type of compound are complex. There are phase changes with variations in temperature and pressure: the conductivity of TTF–TCNQ is 50 times greater at 60 K than at room temperature.

Organic compounds have the attraction of being very diverse because of the enormous number of different molecules that can be used in synthesizing them. As a result, there is always the chance that materials might be found with interesting and worthwhile properties, whereas the possibilities opened up by metals and alloys are very limited and have already been explored. Thus, superconducting organic compounds have been discovered and it is hoped that they will turn out to have high critical temperatures.

There are other one-dimensional conductors with a completely different structure. We might mention polyacetylene or $(CH)_n$ with a formula:

$$
\begin{array}{ccccccc}
 & H & & H & & H & & H & \\
 & | & & | & & | & & | & \\
- & C & = & C & - & C & = & C & -
\end{array}
$$

and polythiazyl $(SN)_n$, formed from a chain of alternating S and N atoms. These materials can have a high conductivity when suitably doped.

Two-dimensional conductors: intercalation compounds of graphite

The graphite crystal is formed from stacked hexagonal layers of carbon (SM, p. 86). Graphite is a conductor, but its conductivity is much greater in the plane of the layers than it is in a direction perpendicular to them.

Graphite has the property of absorbing many atoms or molecules, such as alkali metals, bromine, HNO_3, etc. The absorbed atoms form layers which are interposed (intercalated) between the planes of carbon. Some of these intercalation compounds are very good conductors in the plane of the layers (conductivity comparable with that of copper, about 10 times greater than that of pure graphite). Although the electron mobility in the carbon planes is very high, the number of free electrons in pure graphite is comparatively small. The intercalated layers, on the other hand, are more abundant sources of free electrons and these are easily propagated in the adjacent carbon planes.

3

Magnetic properties of solids

The action of a magnetic field on a solid

What happens when a solid is placed in a magnetic field? An experiment to find out is easy to perform: a specimen merely has to be introduced into a current-carrying solenoid where the magnetic field is both uniform and accurately known from the current flowing in the winding and the number of turns per unit length. The value of the magnetic field near the specimen is then compared with its value before the specimen was introduced.

The result is very clear cut. Almost all substances (solids or liquids, whatever their nature and structure, whether metals, minerals, organic materials, etc) produce only a very small change in the magnetic field at any point in space when they fill the region within the solenoid: the change is no more than about one part in a thousand. In other words, the great majority of solids have virtually no effect on a magnetic field in which they are placed and, to a first approximation, the field has no appreciable effect on the structure or the properties of solids.

However, there are a few exceptional substances, described as **magnetic**, for which the effect of the magnetic field is enormous: the field near the specimen may be increased by a factor of 100 or 1000. Iron, as we all know, is such a substance and, because it is abundant and cheap, many important technological advances have been made possible through the strong magnetic fields that can easily be produced with it. If iron, or any other material as common as iron, did not possess these properties, the present state of our civilization would certainly be very different.

Comparison between the action of magnetic and electric fields

The effect of a magnetic field on matter is markedly different from that of an electric field. First of all, there is no division of materials into two classes similar to conductors and insulators. Secondly, any insulator filling the space between the plates of a parallel-plate capacitor (this is the counter-part of the experiment with the solenoid) always increases its capacitance significantly: it is generally increased by a factor of between 2 and 10. On the other hand, although there are several exceptional solids which have a high relative permittivity (and thus produce large increases in electric fields), their values of ϵ_r in no way approach those of the relative permeabilities of strongly magnetic materials.

There is another difference: strongly magnetic materials can be spontaneous sources of a magnetic field in the absence of an applied magnetic field, forming the **permanent magnets** which are in everyday use. There is no electrical counterpart of comparable importance. There are, of course, **electrets** with a permanent polarization (see p. 65), but these are somewhat out-of-the-ordinary objects whose external effects are always fairly weak. Although some applications of electrets have recently been developed, they remain very marginal compared with those of magnets, which were discovered and exploited many centuries ago, particularly in relation to the compass.

We shall concentrate our attention on those providential exceptions to the natural order of things, the strongly magnetic materials. Our first and main problem here is to explain why they are different from the overwhelming majority of normal substances. We shall then pay particular attention to the properties that are the basis of their most important applications.

The weak magnetic effects mentioned previously, while negligible in practical terms, can nevertheless be measured and investigated if sensitive methods are used. They are theoretically significant since they provide information about the structure of atoms and molecules. Thus, chemists use paramagnetic measurements as a means of identification. However, that would take us outside the scope of this book since it is an atomic or molecular property and not a property of solids as such.

The magnetic moment of the atom

The first condition that must be satisfied if a magnetic field is to exercise a strong influence on a solid is that at least one group of the atoms in it should behave like small magnets, possessing a **magnetic moment**† whose value is a characteristic property of the atom concerned. In addition, a description of the solid from the magnetic point of view includes a specification of the direction of the magnetic moment at each lattice site: the direction may vary from one atom to another even though they may be crystallographically equivalent.

Atomic moments are the basic data as far as magnetism is concerned. It

†A small magnet of moment M is magnetically equivalent to a current loop of area S carrying a current I, such that $IS = M$.

would be possible simply to consider the magnitudes of the various atomic moments as given experimental facts, but it is much more enlightening to relate the moment to the electronic structure of the atom. The relationship is, however, complex and we introduce it in gradual stages.

1. The electron, whether free or bound to an atom, has a spin of $\frac{1}{2}$. This is an inherently quantum mechanical concept for which it would be pointless to seek a classical picture. It means that the electron has an intrinsic angular momentum which can only assume one of the two opposite values $\frac{1}{2}\hbar$ and $-\frac{1}{2}\hbar$ in any given direction, where

$$\hbar = h/2\pi = 1.054 \times 10^{-34}\,\text{J s}.$$

This angular momentum gives the electron a magnetic moment whose magnitude is known as the **Bohr magneton**, μ_B:

$$\mu_B = e\hbar/2m_e = 9.27 \times 10^{-24}\,\text{A m}^2.$$

As in the case of the angular momentum, a measurement of the magnetic moment of the electron along any direction can only yield one of the two opposite values μ_B and $-\mu_B$.

As well as the electrons, of course, an atom contains a nucleus consisting of particles (protons and neutrons) which also have a spin and a corresponding magnetic moment. However, it can be seen from the expression for the Bohr magneton that the mass of the particle occurs in the denominator. It follows that the magnetic moment of the nucleus is negligible compared with that of the atomic electrons.

2. An atomic electron makes a contribution to the magnetic moment of the atom in another way. In its motion around the nucleus, an electron creates a magnetic field just as a small current loop does: the field is that of a magnet whose magnetic moment is related to the *orbital angular momentum* of the electron. The projection of this momentum on to an axis can take on values equal to $m\hbar$, where m is the **magnetic quantum number**. The projection of the magnetic moment on to the same axis is

$$\mu = m\mu_B.$$

In Box 14, we show that this quantum expression is exactly the same as one that can be obtained in an elementary way from the very naive picture of a point-like charged particle moving in a circular orbit around the nucleus.

Box 14

Classical calculation of the magnetic moment of an orbiting electron

We assume that the electron is a point mass m_e with charge e located at a point M in space moving in an orbit around a fixed centre O at which the nucleus is situated.

In addition, it is assumed that the orbit is circular and plane, and that the velocity of the electron in the orbit has a constant magnitude υ.

The classical expression for the angular momentum is

$$\boldsymbol{J} = \overrightarrow{\mathrm{OM}} \times m_e \boldsymbol{v} = m_e R \upsilon \boldsymbol{n}$$

in which R is the radius of the orbit and $\boldsymbol{n}$ a unit vector normal to the plane of the orbit and in a direction such that it forms a right-handed orthogonal system with the velocity and radius vectors.

The orbiting electron is equivalent to a current loop in which the charge $-e$ is passing any point once during the time for one revolution $t = 2\pi R/\upsilon$. It therefore corresponds to a current I given by

$$I = -e/t = -e\upsilon/2\pi R.$$

The magnetic moment of the small loop is then:

$$\mu = IS\boldsymbol{n} = -(e\upsilon/2\pi R)\pi R^2 \boldsymbol{n} = -\tfrac{1}{2}e\upsilon R\boldsymbol{n}.$$

This shows that μ can be written as $\gamma \boldsymbol{J}$, where $\gamma = -e/2m_e$. The quantity γ is known as the **gyromagnetic ratio**. If $J = m\hbar$, we obtain:

$$\mu = -m(e\hbar/2m_e) = -m\mu_B.$$

The negative sign (arising from the negative charge of the electron) can be removed without any inconvenience as was done on p. 138 since, in any applications, a set of m values are involved which are equally likely to be positive and negative, and hence $-m$ and $+m$ are equivalent.

In a classical calculation, for an object of mass M and total charge Q moving in an orbit, the ratio of magnetic moment and angular momentum (γ) is always found to be $Q/2M$ whatever the distribution of charge and mass in the object.

3. We now compare the two expressions for the magnetic moment of the electron, the orbital moment and the spin moment. Both involve the elementary quantity, the Bohr magneton, $\mu_B = e\hbar/2m_e$. The angular momenta in units of $\hbar$ are $\frac{1}{2}$ for the spin and m for the orbital contribution. The formulae for orbital and spin moments can be written in the same way:

$$\mu = g\mu_B\,(\text{angular momentum})/\hbar,$$

the factor g being 2 for spin and 1 for the orbital contribution. The difference between these two values shows that it is no good attempting to explain the intrinsic magnetic moment of the electron in terms of a rotation about its own axis since any classical calculation would give $g = 1$ as for an electron moving in an orbit (see Box 14).

4. The total contribution of an electron to the magnetic moment of an atom is the sum of the two effects, one due to orbital motion and the other due to spin. Complications arise because these two terms are not independent: there is an interaction between them because the magnetic field created by the orbital motion of the electron affects the orientation of its spin moment. Moreover, in the atom as a whole, the contributions from different electrons add together and, again, are not independent of each other.

Overall, the atom has a total angular momentum $J\hbar$, to which there corresponds a total magnetic moment given by the general expression:

$$\mu = g\mu_B J,$$

where g is the **Landé g-factor** (or just the **g-factor**), a quantity depending on the electronic configuration of the atom. If only the spin moments of the electrons were involved, g would be equal to 2; if only orbital moments were involved, g would be equal to 1. In fact, for real atoms, g lies between 1 and 2.

Values of atomic magnetic moments

1 For the free atom

The Pauli principle allows only two electrons with opposite spins to be accommodated in a given orbital. When an orbital is full, the total angular momentum due to the spin of the pair of electrons is zero. In a complete shell or sub-shell, there is an equal number of spins in opposite directions; moreover, the distribution of all the electrons in a complete shell around the nucleus is spherically symmetrical. It can be shown that, in this case, the sum of the orbital momenta is also zero.

A completely filled shell thus contributes nothing to the magnetic moment of the atom. *It is the incomplete shells and 'unpaired' electrons* (i.e. those not matched in opposing pairs) which are the sources of the atomic magnetic moment.

We can now see why the magnetic moments of atoms always have a magnitude equal to a small number of magnetons only. It is because, in an atom with many electrons, most of them are located in the inner complete shells and therefore contribute nothing to the moment. In particular, in atoms where all the shells or sub-shells are full, as in the inert gases, the atomic moment is zero.

We can also see why the strongest magnetic moments are to be found among the transition elements, since these are characterized by the existence of an incomplete shell inside the outer shell.

2 *For an atom in a solid*

The situation is simple for ionic crystals: the constituent ions have an electron cloud that is highly localized around the nucleus and which ends in a completely filled outer shell. If ions have a non-zero magnetic moment, they have an incomplete *inner* shell.

With metals, the model of a moment attached to each lattice site in the crystal is not completely satisfactory since, as well as the electrons localized in the inner electrons shells, there are also the conduction electrons, non-localized and subject to quantum physics alone. A theory of the magnetism of metals has been created from band theory but, in general, we shall restrict ourselves to the idea of a moment localized in the atom. However, it should be pointed out that for metals, because of the non-localized nature of the free electrons, the atomic moments are not integral multiples of the Bohr magneton, e.g. 2.22 for iron, 0.6 for nickel.

There is quite a subtle point which needs clarification when we are discussing the magnetic moment of atoms. We treat it as an ordinary vector, of the same nature as the angular momentum of the atom. The latter is, however, a quantum property and does not have the same characteristics as a classical vector based on classical geometry. Thus, the angular momentum $J\hbar$ has a magnitude $\sqrt{J(J+1)}\hbar$ and its projection along any direction can take on the values $m_J\hbar$, where m_J is one of the $2J+1$ values $-J, -J+1, \ldots$, up to J. This is unlike the classical vector μ, whose projection varies continuously from $-\mu$ to $+\mu$ depending on the angle it makes with the direction concerned. If the individual moments in the crystal have a random orientation, their resultant projection on any axis is zero. In quantum theory, this means that all the values of m_J are equally probable.

In practice, particularly if J is large, the difference between the two representations is not very great and this is our justification for treating the magnetic moment like an ordinary vector.

Diamagnetism and paramagnetism

Atoms with zero magnetic moment: diamagnetism

The magnetic field created by the solid in this case can only be zero because the atoms have no resultant magnetic moment. It is also true that an externally applied magnetic field has virtually no effect on the solid. However, there is a residual effect due to electromagnetic induction.

It is well known that, when a magnetic field is established near a conducting circuit, the free electrons in the metal are set in motion: there is an induced current flowing in such a direction that the magnetic field it produces opposes the inducing field (Lenz's law).

In an atom subjected to a magnetic field, the electron motion is modified by the force due to electromagnetic induction. The effect is not transient as it is in a metal with its electrical resistance, but is permanent since there is no decay of the electronic motion within the atom. Thus, an atom with zero magnetic moment acquires a moment under the influence of the field, but in a direction opposite to that of the applied field since it is the effect of electromagnetic induction. A calculation shows that the moment is extremely small.

This phenomenon is called **diamagnetism** and it occurs in all atoms. However, when the atom *does* have a magnetic moment, the diamagnetism is masked by a greater effect which we look at in the next section. In practice, therefore, this very weak magnetic effect can be neglected and we merely mention it for completeness.

Atoms with a non-zero magnetic moment: paramagnetism

Each atom experiences the force exerted by an external magnetic field on its atomic moment. However, we should first of all dismiss an idea suggested by our experience of macroscopic objects but which is not accurate on an atomic scale: that the effect of the force exerted by the field would be to align the atomic moment along the field direction. A small magnetized needle behaves in that way, but this is because of damping due to friction. If the needle were perfectly free to move around its pivot, it would oscillate indefinitely around the direction of the applied field. However, atomic motion is not damped and in fact, by the gyroscopic effect, the atomic moment moves around the field direction at a uniform

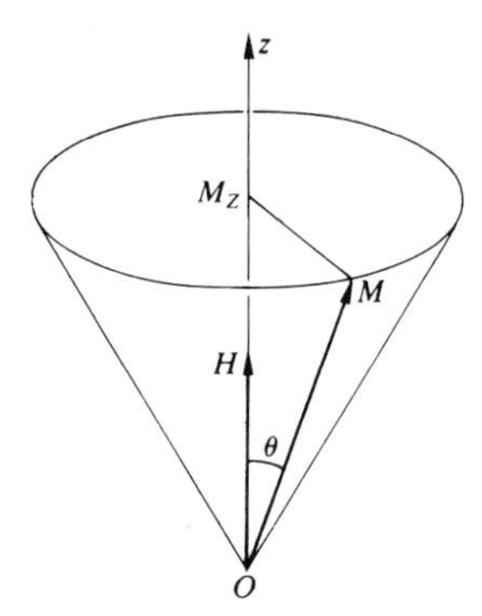

Fig. 3.1. Precession of the atomic magnetic moment around the direction of an applied field.

speed and making constant angle with this direction. This type of motion is called **precession** and, while it is taking place, the projection of the moment on the field direction remains constant (Fig. 3.1).

According to the quantum picture, this projection can take one of the values $g\mu_B m_J$, the number m_J varying in steps of 1 from J to $-J$. The energy of the magnetic moment in the field† is:

$$E = -g\mu_B m_J \mu_0 H$$

The precession induced by the action of the field on the atomic dipole leaves the quantum number m_J constant. But the energy of the atom may vary: as these variations occur, the atomic moment may jump from one m_J value to another, i.e. it may change its orientation with respect to the field. What makes these variations in orientation occur? There are two opposing effects: one is the tendency of m_J to be as large as possible so as to reduce the magnetic energy to a minimum; the other is thermal agitation which favours disorder, i.e. a uniform population of all the m_J levels from $-J$ to $+J$.

In this system, as in others we have encountered (p. 4), equilibrium is defined by Boltzmann's law: the probability that a level m_J is occupied is proportional to the exponential factor:

$$\exp\left(-\frac{E}{k_B T}\right) = \exp\left(g\,\mu_B m_J \frac{\mu_0 H}{k_B T}\right).$$

Let us specify some orders of magnitude for a real everyday situation. For example, H is a field easily obtained with an ordinary solenoid ($10^3\,\mathrm{A\,m^{-1}}$ or $\mu_0 H \approx 10^{-3}\,\mathrm{T}$). T is room temperature. We shall take $g = 2$ and a system where there are two possible values for m_J: $+\frac{1}{2}$ and $-\frac{1}{2}$. We then find that, between the levels with the lower and higher energies, the difference in occupation is only 1/1000 in relative values. The distribution of moments over the two levels is thus very nearly uniform. The magnetization of the solid as a whole is therefore very weak. The situation would be the same for systems with many m_J levels.

This is the explanation of the fact which we mentioned at the beginning of the chapter: that the magnetic field has in general only a small effect on the solid. If its atoms have magnetic moments, they are only very partially oriented by the applied magnetic field: this is **paramagnetism**. Although paramagnetic effects can be neglected in practical terms, we are going to look at them in more detail because they form an essential introduction to the understanding of the properties of strongly magnetized materials.

†We shall not introduce the magnetic induction vector $\boldsymbol{B}$ which, outside magnetized material is given by $\boldsymbol{B} = \mu_0 \boldsymbol{H}$. Our aim is not to deal with the physics of magnetic fields, but to describe the relationships between magnetization and the structure of matter by considering simple and easily appreciated situations (e.g. Fig. 3.15).

The magnetization of a solid is specified by the quantity M, the intensity of magnetization or simply the magnetization, defined as its magnetic moment per unit volume, i.e. the 'mean' atomic moment multiplied by the number of atoms per unit volume. A calculation using Boltzmann's law (Box 15) shows that, for fields that are not too strong and temperatures that are not too low, M is proportional to H. The ratio, known as the **magnetic susceptibility**, χ_m, is typically of the order of 10^{-3}. This is the paramagnetic state. The susceptibility is inversely proportional to the temperature (Curie's law†):

$$\chi_m = M/H = C/T$$

where C is the Curie constant.

However, in a strong field and when the atomic moment is large and the temperature very low, Curie's law is no longer valid: we approach **saturation**, where all the moments are aligned (see Box 15). Thus, at 1 K and in a field of $3 \times 10^6\,\mathrm{A\,m^{-1}}$, gadolinium sulphate has a magnetization of $3 \times 10^5\,\mathrm{A\,m^{-1}}$. This is the first example we have met of a solid that can be appreciably magnetized, and even here the solid has to be placed in extreme conditions.

Box 15

Calculation of the magnetization as a function of temperature and applied field

Consider a body consisting of identical atoms, characterized by the quantum number J placed entirely within a uniform magnetic field $\boldsymbol{H}$. If the field lies along the z-axis, the energy of a single atomic moment μ in the field is:

$$E = -\mu_0 \mu \cdot \boldsymbol{H} = -\mu_0 \mu_z H$$

where the projection μ_z of the atomic magnetic moment along the direction of the field can only take the following $2J+1$ discrete values:

$$\mu = m_J g \mu_B$$

with $m_J = -J, -J+1, \ldots, J$.

It follows that the energy of the magnetic moment is quantized:

$$E = -m_J g \mu_B \mu_0 H.$$

†What is in fact observed is the paramagnetic susceptibility minus the diagmagnetic susceptibility, the latter existing in all atoms whether or not they have a magnetic moment. The diamagnetism is, however, completely negligible in relative terms.

The quantum state defined by m_J exists at a temperature T with a probability proportional to the Boltzmann factor $\exp(-E/k_B T)$.

The probability of finding the projection of the magnetic moment along Oz being equal to $m_J g \mu_B$ is therefore:

$$\frac{1}{Z}\exp\left(m_J g \mu_B \frac{\mu_0 H}{k_B T}\right)$$

where Z is introduced to normalize the probabilities:

$$Z = \sum_{m_J=-J}^{J} \exp\left(m_J g \mu_B \frac{\mu_0 H}{k_B T}\right)$$

When the magnetic field is not zero, there is a greater probability that m_J is positive than there is that it is negative. This means that, at temperature T, the magnetic moment has on the average a positive component along the field H. This mean value becomes greater as H increases or as T falls, and is calculated in the usual way (see Boxes 1 and 2) by multiplying each possible value of the projection $m_J g \mu_B$ by the probability of its existence, and then summing over all the possible values of m_J:

$$\langle\mu\rangle = \frac{1}{Z}\sum_{m_J=-J}^{J} m_J g \mu_B \exp\left(m_J g \mu_B \frac{\mu_0 H}{k_B T}\right)$$

The rest of the calculation, although posing no great problems, is nevertheless quite tedious and laborious. It therefore seems preferable to give a detailed calculation just for the simplest case in which the magnetic moments are due to spin momentum only, i.e. $J = \frac{1}{2}$, $g = 2$ and where m_J can only take the values $+\frac{1}{2}$ and $-\frac{1}{2}$. In this case, the expressions for Z and $\langle\mu\rangle$ become

$$Z = \exp\left(\frac{\mu_B \mu_0 H}{k_B T}\right) + \exp\left(-\frac{\mu_B \mu_0 H}{k_B T}\right)$$

$$\langle\mu\rangle = \frac{\mu_B}{Z}\left[\exp\left(\frac{\mu_B \mu_0 H}{k_B T}\right) - \exp\left(-\frac{\mu_B \mu_0 H}{k_B T}\right)\right]$$

i.e.

$$\langle\mu\rangle = \mu_B \tanh(\mu_B \mu_0 H/k_B T),$$

where $\tanh x$ is the hyperbolic tangent with the value $(e^x - e^{-x})/(e^x + e^{-x})$.

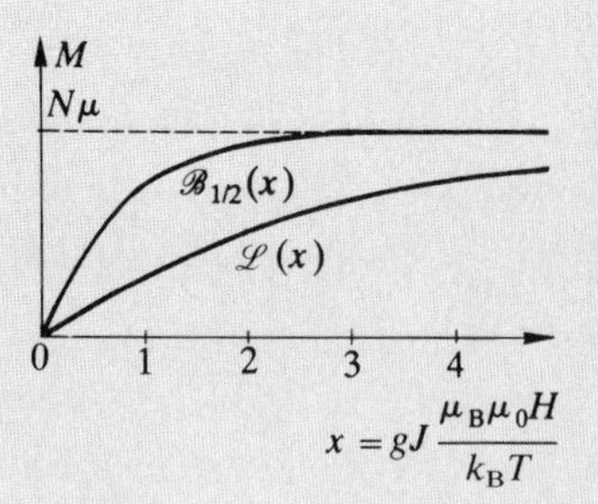

Fig. 3.2. Intensity of magnetization and its variation with $x = gJ\mu_B\mu_0 H/k_B T$ calculated for the Brillouin function $\mathscr{B}_{\frac{1}{2}}$ (for $J = \frac{1}{2}$, $g = 2$) and for the Langevin function $\mathscr{L}$, the limit of the Brillouin function as J tends to infinity.

We then obtain the magnetization by multiplying $\langle\mu\rangle$ by the number of atoms per unit volume, N:

$$M = N\mu_B \tanh(\mu_B\mu_0 H/k_B T).$$

The curve giving M as a function of $x = \mu_B\mu_0 H/k_B T$ is plotted in Fig. 3.2. When x tends to infinity, i.e. for strong fields or low temperatures, the orientation in the field is perfect and M becomes equal to $N\mu_B$. On the other hand, when H is weak and/or when T is high, the hyperbolic tangent can be put equal to its argument x, and we then obtain

$$M = N\frac{\mu_B^2\mu_0}{k_B}\frac{H}{T}.$$

The magnetization M is proportional to H and the constant of proportionality, or magnetic susceptibility χ_m, is given by:

$$\chi_m = N\frac{\mu_B^2\mu_0}{k_B}\frac{1}{T}.$$

The susceptibility varies inversely with temperature, which is **Curie's law**. We emphasize that the law is only valid if

$$\mu_B\mu_0 H \ll k_B T.$$

However, the range of validity is very wide. Even for strong fields ($\mu_0 H \approx 1\text{T}$), the temperature has to fall to a few K to cause Curie's law to fail.

In the general case, where J is different from $\frac{1}{2}$, the calculation gives the following result:

$$M = NgJ\mu_B \mathscr{B}_J(x)$$

where $x = gJ\mu_B\mu_0 H/k_B T$, and where $\mathscr{B}_J$, the **Brillouin function**, is given by the following expression:

$$\mathscr{B}_J(x) = \frac{2J+1}{2J} \coth\left(\frac{2J+1}{2J} x\right) - \frac{1}{2J} \coth \frac{x}{2J}$$

This function, which reduces to the hyperbolic tangent for $J = \frac{1}{2}$, behaves qualitatively in a similar way to the tanh function. It has a limiting value of 1 towards infinity, which describes **saturation**, and near zero it is linear:

$$\mathscr{B}_J(x) \sim \frac{1}{3}\frac{J+1}{J} x$$

which gives Curie's law $\chi_m = C/T$, with

$$C = \frac{N\mu_{\text{eff}}^2 \mu_0}{3k_B} \quad \text{and} \quad \mu_{\text{eff}} = \sqrt{J(J+1)}g\,\mu_B.$$

A particularly important value is the 'classical limit', obtained for large quantum numbers. When J tends to infinity, the Brillouin function simply converges to a limiting function, the **Langevin function**, given by

$$\mathscr{L}(x) = \coth x - \frac{1}{x}.$$

The curves for $\mathscr{B}_{\frac{1}{2}}(x) = \tanh x$ and $\mathscr{L}(x) = \mathscr{B}_\infty(x)$ are given in Fig. 3.2.

Ferromagnetism

We now come to the exceptional cases of substances which can be strongly magnetized under ordinary conditions. Among the thousands of natural minerals in the Earth's crust, there is one, an oxide of iron, Fe_3O_4, known as magnetite, which can be strongly and spontaneously magnetized: this has been known since antiquity in the form of 'lodestone'. Among the metals, there is one, iron, which can be strongly magnetized at ordinary temperatures in a current-carrying coil, and two others, nickel and cobalt, which can be magnetized a little less strongly. After a considerable amount of research into magnetism and magnetic materials, the number and diversity of 'magnetic substances' has in fact been somewhat extended, but in spite of that they remain rare exceptions.

The physicist's task is to discover the origin of the extraordinary properties of these materials, while attempting to retain the model incorporating the atomic carriers of magnetic moment that is valid for non-magnetic substances.

The molecular field

The great majority of solids are magnetically inactive. This is not because the individual atomic magnetic moments are too weak, but because the moments are randomly oriented and remain so even in an external field which, however intense, has only a small effect.

The condition for strong magnetization is that the atomic moments become *globally* ordered. This is not normally possible because the action of the field tending to align the moments is swamped by the disordering effect of thermal agitation. We have mentioned that better alignment can be achieved by increasing the field strength or lowering the temperature. Yet iron is magnetized at ordinary temperatures and in fields that are weak, or even zero. Under these conditions, the external field is not capable of producing sufficient alignment of the atomic moments on its own, so what is it that does?

In 1910, Pierre Weiss responded to this question by suggesting an idea which, with modifications and refinements, still remains the basis for the theory of how iron behaves in a magnetic field and, more generally, for ferromagnetism.

In this theory, the applied field is assisted by a much stronger magnetic field created by the iron itself, which Weiss called the **molecular field**.[†] Without attempting to discover the origin of such a field, Weiss first showed what could be deduced from such a hypothesis.

The existence of the molecular field at a site is due to the action of neighbouring atoms as they begin to align themselves in the direction of the applied field, and thus when the magnetization is not zero. Weiss then simply assumed that the molecular field parallel to the common direction

[†]This name has been kept even though it appears unsuitable today since there are no molecules in metallic iron. A more correct nomenclature, now used in statistical mechanics, is the **mean field approximation** (MFA).

of $\boldsymbol{M}$ and $\boldsymbol{H}$ is proportional to the magnetization M. The *effective* field acting inside the solid is therefore of the form $H + \lambda M$.

Suppose in addition that the action of the effective field is determined by Curie's law, which takes into account the action of the external field on a paramagnetic. We can then write:

$$\frac{M}{H + \lambda M} = \frac{C}{T}$$

which can be put into the form:

$$\frac{M}{H} = \frac{C}{T - \lambda C} = \frac{C}{T - T_c}.$$

This is the **Curie–Weiss law** in which a new parameter, λC, appears, with the dimensions of temperature and denoted by T_C.

When $T = T_C$, the susceptibility becomes infinite. This means that at the temperature T_C, the magnetization may be different from zero, even for an external field that tends to zero. Physically, this describes the possibility of **spontaneous magnetization**, which is exactly what occurs in a ferromagnetic below a critical temperature known as the **Curie point**. The Curie point for iron, for example, is $T_C = 1043\,\text{K} \approx 770°\text{C}$.

When $T > T_C$, iron behaves more or less like a paramagnetic: the magnetization remains very small and the susceptibility varies with temperature as predicted by the Curie–Weiss law.

Below T_C, M can be very high and we are no longer in the region where M is proportional to H, so that the Curie–Weiss law has no meaning. We must use the real $M = f(H)$ curve (p. 146), but here, the effective field is $H + \lambda M$. In fact, we can take this field as equal to λM, since we wish to calculate the spontaneous magnetization, i.e. in zero applied field. The result of such a calculation, given in Box 16, is an expression giving the sponteneous magnetization as a function of temperature (Fig. 3.4).

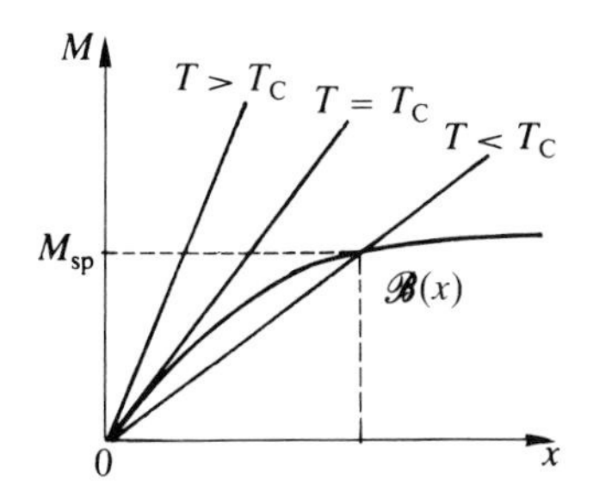

Fig. 3.3. The calculated spontaneous magnetization and its variation with temperature.

Box 16

Calculation of spontaneous magnetization using the molecular field approximation

In the molecular field approximation of Weiss, it is as if the individual magnetic moments were subjected to an external field H supplemented by another field λM proportional to the magnetization. The magnetization can then be calculated as in Box 15, replacing H by $H + \lambda M$. In the general case of magnetic moments characterized by the quantum number J, we therefore have:

$$M = NgJ\mu_B \mathscr{B}_J(x)$$

with

$$x = \frac{gJ\mu_B\mu_0}{k_B T}(H + \lambda M).$$

It is then a question of solving this equation for M, as an unknown appearing on both sides. We shall use a graphical method for the case of zero applied field ($H = 0$), when we have to solve the system of equations:

$$\begin{cases} M = NgJ\mu_B \mathscr{B}_J(x) \\ x = \dfrac{g\lambda J\mu_B\mu_0}{k_B T} M. \end{cases}$$

By using the expression for the Curie constant:

$$C = \frac{NJ(J+1)(g\,\mu_B)^2\,\mu_0}{3\,k_B}$$

and that for the critical temperature $T_C = \lambda C$, we can not write the system of equations in the form:

$$\begin{cases} \dfrac{M}{NgJ\mu_B} = \mathscr{B}_J(x) \\ \dfrac{M}{NgJ\mu_B} = \dfrac{J+1}{3J}\dfrac{T}{T_C}\,x. \end{cases}$$

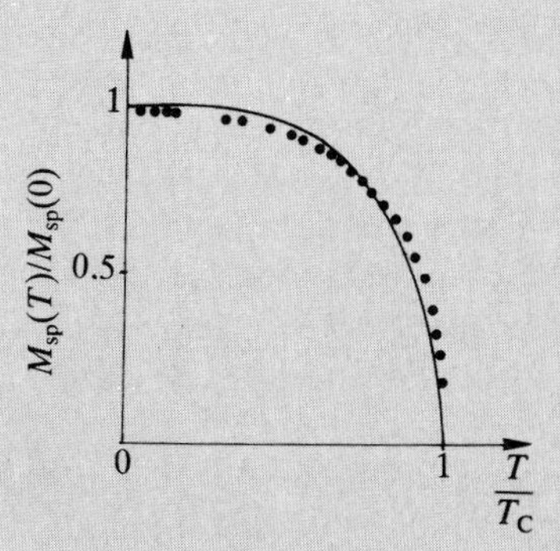

Fig. 3.4. Comparison of the theoretical and experimental values of the spontaneous magnetization in nickel: T_C = 631 K.

We plot both the variation of the Brillouin function $\mathcal{B}_J$ and that of the linear function

$$x \to \frac{J+1}{3J} \frac{T}{T_C} x$$

in the same figure (Fig. 3.3). The value of the vertical coordinate at the point of intersection of these two curves gives the value of $M/NgJ\mu_B$.

The slope of the Brillouin function at the origin is $(J+1)/3J$. It can be seen that, if $T > T_C$, the only solution is $M = 0$. Above T_C, the iron is therefore unmagnetized in the absence of an external field and it behaves like a paramagnetic.

On the other hand, when $T < T_C$, the straight line cuts the curve at a point away from the origin, giving rise to a non-zero magnetization M_{sp}. A consideration of the energy of the system would show that it is lower for this solution than for $M = 0$. As a result, the iron naturally possesses a non-zero spontaneous macroscopic magnetization M_{sp} in the absence of an applied field.

This graphical solution yields M_{sp} as a function of T, and the resultant curve is shown in Fig. 3.4. The magnetization M_{sp} falls from its maximum value M_s (saturation), $NgJ\mu_B$, and tends to zero at $T = T_C$. The behaviour of M_{sp} near T_C is obtained by expanding the Brillouin function to the third order term in x. This shows that M_{sp} varies as $(T_C - T)^{1/2}$ near T_C.

The molecular field approximation, used more generally under the name of the mean field, ignores fluctuations of the atomic magnetic moments around their mean values: it is assumed that the 'molecular' field at any atomic site is exactly λM.

This approximation would undoubtedly not be justified for a one-dimensional chain of atoms. (In that case, it is found that there can be no spontaneous magnetization as soon as the temperature differs from zero.) On the other hand, for a two-dimensional plane of atoms, or in the more realistic case of a three-dimensional crystal, the approximation does enable the existence of spontaneous magnetization to be found below a characteristic temperature T_C.

In theories more advanced than the mean field theory, it can be shown that M_{sp} approaches zero at T_C like $(T_C - T)^{\beta}$. The value of β, known as the **critical exponent,** is not in general equal to $\frac{1}{2}$, but depends on the symmetry of the interactions between magnetic moments and on the dimension of space.

The renormalization group theory of Kadanoff, Fisher, and Wilson (Nobel Prize winner, 1982) has shown that the mean field theory becomes exact in the abstract case of spaces of more than four dimensions. This theory has also enabled the critical exponents to be calculated with great accuracy.

To compare theory with experiment, the spontaneous magnetization must be measured as a function of temperature. It is only after we have made a more thorough study of ferromagnetism that we can show that the spontaneous magnetization M_{sp} is represented by the magnetization of iron at saturation, i.e. the value towards which it tends as the field applied to the iron is increased (see p. 164).

Weiss's idea, even though its application involves only simple calculations, accounts satisfactorily for the ferromagnetic state and for its upper limit at the Curie point. Clearly, we must now tackle the problem of the origin of the molecular field, but before doing that we offer two comments from a phenomenological point of view.

1. According to Fig. 3.4, the ferromagnetic–paramagnetic transition is very sharp. This is a characteristic of a **cooperative phenomenon**. The aligning action of the molecular field becomes stronger as the atomic moments themselves become more aligned. Starting from a crystal with completely aligned moments at a very low temperature, we raise the temperature gradually. While it is still low, the magnetization remains constant, i.e. the moments remain aligned because of their interaction and in spite of thermal agitation. Then a few moments become misaligned: the action of the molecular field on each atom is weakened, which causes a misalignment of some more moments. The interactions become even weaker, the disorder grows, weakening the interactions further, and so on. The process is self-accelerating: there is no slow and gradual decrease in

the alignment of the moments. As the critical temperature is approached, the situation tends to become catastrophic and the order disappears abruptly.

This is characteristic of cooperative phenomena, and many examples with similar features are known in physical transformations: they are second order phase transitions.

2. The Weiss theory is one example of a method of approximation known as the **mean field** approximation. An atom interacts mainly with its nearest neighbours and it is assumed that the interaction is the same as if all the neighbours were in the average state of the complete atomic assembly. This is an approximation that makes the calculation easier but it neglects fluctuations in local order. A refined version of the theory takes these into account.

What is the molecular field?

Although it appears obvious that the assumption of the molecular field is the key to the explanation of ferromagnetism, we cannot remain content to postulate its existence without asking what physical phenomenon is the cause of it.

We first reject one possible origin of the effect, even though it seems the most natural one: this is the action of classical magnetic forces between neighbouring atomic moments. Our rejection is based on the following argument.

The value of the molecular field can be evaluated for a given substance by using the relationship $T_C = \lambda C$. Since the Curie point T_C is known and the paramagnetic constant C can be deduced from the atomic magnetic moment (see p. 147), the value of λ can be calculated and we find that for iron $\lambda = 5000$. The magnetization of iron at saturation is $M = 2 \times 10^6\,\mathrm{A\,m^{-1}}$, so that the molecular field λM would be of the order of $10^{10}\,\mathrm{A\,m^{-1}}$. But the field created at a lattice site by the moments of neighbouring atoms, assumed to be all parallel, is less than a thousandth of this value. The origin of the molecular field postulated by Weiss must lie elsewhere.

Classical physics proves powerless to provide a solution to the problem. Quantum mechanics, however, has been able to do so by predicting the existence of *coupling forces* or *interactions* between the magnetic moments of neighbouring atoms which cause the moments to align themselves preferentially in directions parallel to each other. These coupling forces, of purely quantum origin, differ in type according to the atomic structure and the type of chemical bond involved. One contribution is the 'exchange' term (Heisenberg) in the calculated energy of the system. This term arises from the indistinguishability of electrons: an interchange of two electrons must not change either the wave function or the energy. The exchange energy always tends to make the magnetic moments of neighbouring atoms align themselves parallel to each other but in opposite directions. This term, which is predominant in some insulators, only plays a small part in metals since the delocalization of the valence electrons necessitates the introduction of other terms. One of these, known as the Coulomb term, encourages the spins to align themselves in the same direction because of the Pauli principle. The coupling results from all these effects and from others that are more complex. The interatomic distance, the more or less strong delocalization of the electrons, the degree to which the electron shell responsible for magnetism is filled, all these are factors determining the sign and the value of the total coupling between the magnetic moments of neighbouring atoms.

To attempt an explanation of how such a coupling follows from quantum theory, we take the simple example of the hydrogen molecule (SM, p. 16), which can be calculated. The hydrogen molecule consists of two H^+ nuclei separated by 0.108 nm and two electrons. Starting from Schrödinger's equation, a quantum mechanical calculation yields the mean distribution of the electrons or, what amounts to the same thing, the mean distribution of the negative charge density around the nuclei. The binding energy between the atoms can also be found: it is the difference between the energy of the H_2 molecule and the two separated H atoms. The spatial distribution of the charge and the binding energy of the molecule are different for the two cases in which the electron spins are parallel or antiparallel. The binding energy is positive if the spins are parallel and the molecule then tends to split spontaneously into two atoms; if the spins are antiparallel, the binding energy is negative: the two atoms are then attracted to each other to form the molecule, as if there is a force between them, the coupling force. The energy gained when the nuclei are separated by their equilibrium distance (0.108 nm) is 4.75 eV. This is electrostatic energy, since it arises from a modification in the distribution of the negative charge. It is of purely quantum origin, since in classical theory the position of the electrons is not influenced by their spins, while the orientation of the latter is involved in quantum theory because of the Pauli principle.

If the coupling energy were attributed to a 'molecular' magnetic field H_m, the work needed to reverse a magnetic moment would be $2\mu_B\mu_0 H_m$. In the H_2 molecule, H_m would be $3 \times 10^{10}\,A\,m^{-1}$. This is a very high value and is of the same order as that found in iron.

We now pass on to the case of an iron atom in a crystal. The large number of neighbours and the complexity of the atom with so many electrons makes it impossible to carry out exact calculations. But the facts revealed by the H_2 molecule explain qualitatively what happens in iron. The electrons in the incomplete sub-shell, which are responsible for the magnetism of iron, have distributions which, for a pair of neighbours, partially overlap each other. The electron distribution therefore depends on the orientation of the spins, just as the atomic binding energy does.

It is found that, in metallic iron, the energy of atoms with parallel magnetic moments is lower than that when they are antiparallel. This is what produces the tendency for the moments to align themselves in the same direction, which is the basic phenomenon of ferromagnetism.

As in the simple hydrogen molecule, the effect is of quantum origin and is due to electrostatic interaction between electrons. However, the interaction may be formally expressed as a magnetic interaction through the medium of an internal magnetic field due to the atomic moments: this is the mysterious molecular field that Weiss postulated a priori.

The ferromagnetic elements

Quantum theory provides a rational basis for the Weiss conjecture: it explains why the number of ferromagnetic substances at ordinary temperatures in so small. The first condition to be satisfied is the existence of an appreciable atomic magnetic moment. As we have seen, few electronic configurations permit this, so that there are not many 'candidates' for ferromagnetism. Secondly, the coupling energy must have such a sign that it tends to make the moments parallel. Finally, this energy must be great enough compared with the energy of thermal agitation $k_B T$ for the alignment of moments to persist up to temperatures above room temperature.

Table 3.1. Curie point and spontaneous magnetization at 0 K for some ferromagnetics

Substance	Curie point (K)	Atomic magnetic moment (magnetons)	Spontaneous intensity of magnetization at 0 K ($10^6\,A\,m^{-1}$)
Iron	1043	2.2	1.7
Cobalt	1400	1.7	1.4
Nickel	631	0.6	0.51
Gadolinium	292	7.1	2.0
Dysprosium	85	10.1	2.9
Cu_2MnAl	710	3.5	0.50
MnBi	630	3.5	0.68
CrTe	339	2.5	0.25
CrO_2	392	2.0	0.51
EuO	69	6.8	1.9

According to Table 3.1, there are only three ferromagnetic metals at room temperature: iron, nickel, and cobalt. Two others, dysprosium and gadolinium, have large moments but are only ferromagnetic below 85 K and 292 K respectively. Manganese has quite a large magnetic moment, larger than that of iron, but the interaction energy does not produce alignment of moments. However, if manganese atoms are incorporated in the crystal structure of an alloy along with two other non-magnetic elements, copper and aluminium, the conditions are such that the magnetic interaction aligns the moments of the manganese. This forms a Heusler alloy, Cu_2MnAl, which is ferromagnetic up to 710 K.

Other types of magnetic ordering in crystals

Crystals are now known in which the interaction between magnetic moments produces types of ordering different from that in the simple case of ferromagnetism. These are revealed by unusual magnetic properties: either the magnetism is weak but the behaviour is different from that of normal paramagnetics, or else there is a strong magnetization comparable with that of ferromagnetics. The magnetic structures described below were first proposed in order to account for these macroscopic properties: Néel (Nobel Prize winner, 1970) discovered antiferromagnetic order in this way. Today, however, we can use neutron diffraction to map out the orientation of the magnetic moments in the unit cell of a crystal directly. This is possible because the amplitude scattered by an atom depends on the direction of its magnetic moment in relation to that of the neutron.

Antiferromagnetism

Consider the compound MnO. Above 122 K, X-rays and neutrons give the same diffraction pattern, from which the structure shown in Fig. 3.5(a) is deduced. Below 122 K, a critical temperature known as the Néel temperature, neutron diffraction shows that the magnetic moments of the Mn^{2+} ions have regularly alternating directions in successive reticular planes as in Fig. 3.5(b). The magnetic interaction favours pairing between neighbours with antiparallel moments, giving rise to the antiferromagnetic structure.

In the absence of an applied field, the magnetic moment of the crystal as a whole is zero, both in the ordered and disordered state. An antiferromagnetic material never becomes strongly magnetized, but its behaviour when a field is applied is unusual (Fig. 3.6) and this is what drew attention to MnO in the first place.

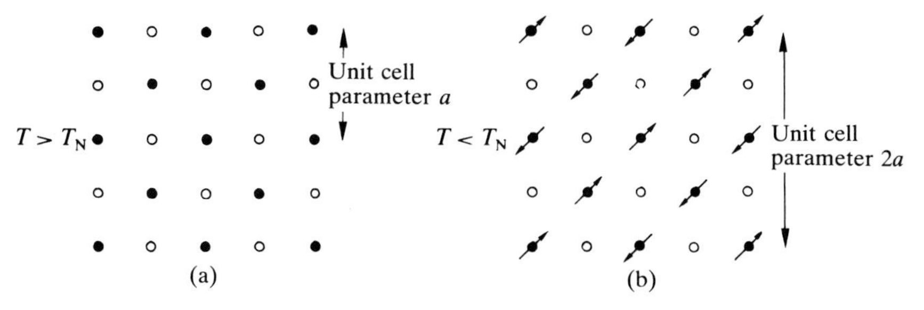

Fig. 3.5. The arrangement of Mn^+ ions (●) and O^- ions (○) in a basal plane of the cubic unit cell of MnO. (a) When $T > T_N$, the magnetic moments in zero field are disordered. (b) When $T < T_N$, the moments alternate regularly from one plane to the next and this doubles the size of the unit cell as revealed by neutron diffraction.

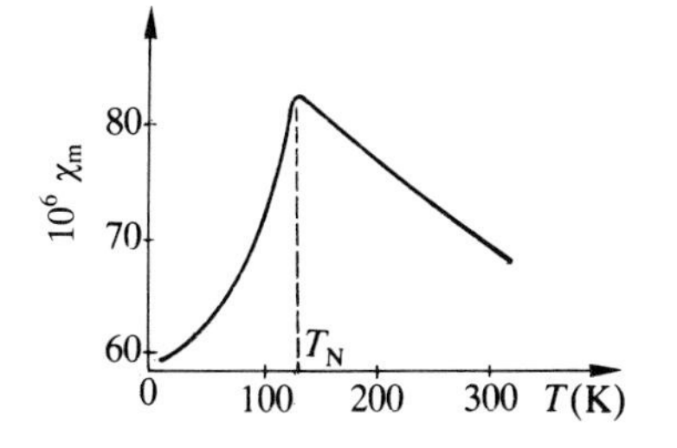

Fig. 3.6. The variation of the magnetic susceptibility of MnO with temperature.

Helimagnetism

Much more complex types of ordering have been revealed by neutron diffraction. Thus, in the alloy $MnAu_2$, the Mn atoms are arranged in identical parallel planes separated by a distance d. In each plane, the magnetic moments are all parallel, but in passing from one plane to the next their direction turns through a constant angle: 51° in this case. The whole crystal has a zero moment. For a given column of Mn atoms, the moments form a helix (Fig. 3.7) with a pitch that is not a simple fraction of d (360/51 = 7.06), i.e. the crystal unit cell and the pitch of the helix are **incommensurable**. This is an example of a phenomenon that has been observed in various forms in certain crystals and is currently the subject of a great deal of research.

To explain such complications in the ordering of the magnetic atoms, the concept of **frustration** is introduced. Assume that there is an interaction not only between the magnetic moments of nearest neighbour atoms, but also between those of next nearest neighbours. Suppose also that nearest neighbours have a tendency to align themselves parallel to each other, while next nearest neighbours have a tendency to become aligned antiparallel to each other. In a group of three adjacent atoms, 2 will align itself parallel to 1, and 3 parallel to 2, but the most favourable direction for 3 in

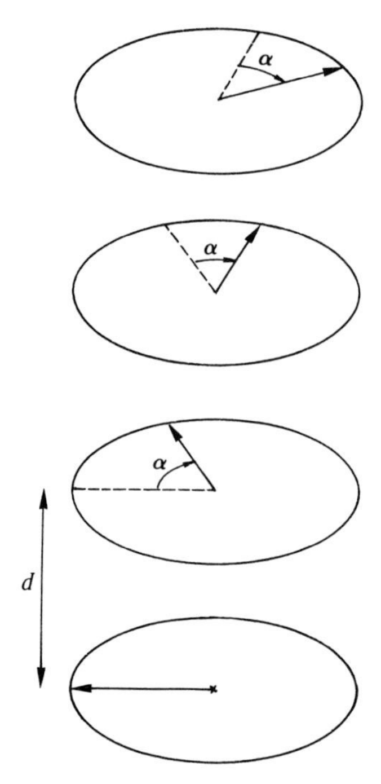

Fig. 3.7. Helical magnetism: the helical arrangement of the magnetic moments in a column of ions.

relation to 1 is antiparallel, which conflicts with its tendency to be parallel to 2. Atom 3 is said to be in a frustrated state. The situation is alleviated by the adoption of a helical structure with an angular displacement between successive moments of 51°. This makes the interaction between nearest neighbours slightly positive and that between next nearest neighbours (with a relative angular displacement of 102° and thus greater than 90°) slightly negative.

Ferrimagnetism†

Ferrimagnetics are important materials, strongly magnetic and with many applications. Unlike the substances considered so far, they contain two types of ion with different magnetic moments. One example is that of the earliest known natural magnet, magnetite, which has a chemical formula Fe_3O_4 (or $FeO.Fe_2O_3$), i.e. two Fe^{3+} ions and one Fe^{2+} ion per molecule. The Fe^{3+} and Fe^{2+} ions have magnetic moments of 5 and 4 magnetons respectively. The relatively bulky oxygen ions form a close-packed cubic

†No chemical significance should be attached to the prefixes ferro- and ferri-. In this case, they are not used to denote the valence states of the iron ion.

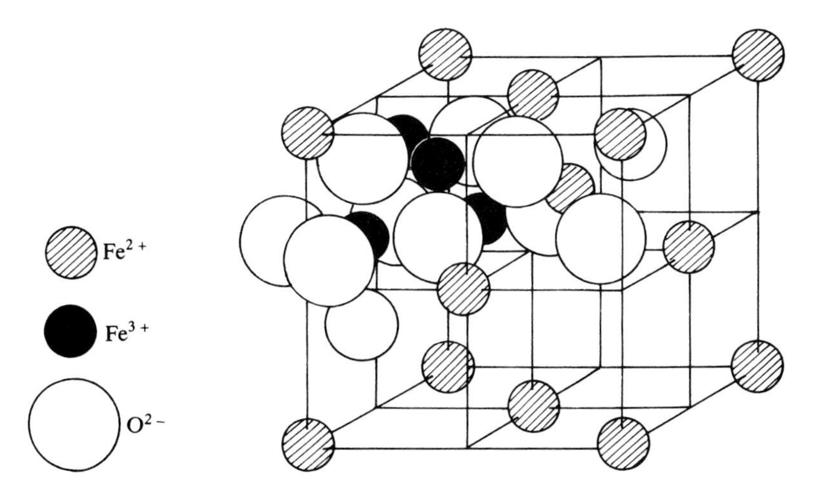

Fig. 3.8. The unit cell of Fe_3O_4: the spinel structure.

lattice (Fig. 3.8) in which there are two types of interstitial site, both accommodating iron ions but with different geometries: one type of site is tetrahedral and the other octahedral. The magnetic ions are thus distributed over two **sub-lattices:** in one, there is one Fe^{3+} ion per 'molecule' and in the other, there is one Fe^{3+} and one Fe^{2+} ion. In each sub-lattice, the exchange interactions align the moments parallel to each other, but the essential point is that the orientations in the two sub-lattices are antiparallel. Their contributions to the total moment must therefore be subtracted from each other: the moment of the crystal per molecule is thus $(5 + 4) - 5 = 4$ magnetons, and this is confirmed by experiment. As with ferromagnetics, the ordering of the moments disappears sharply at a critical temperature also called the Curie point (Table 3.2).

Magnetite is representative of a whole series of ferrimagnetics known as **ferrites**. Their general formula is (Fe_2O_3.MO) where M is a cation such as Zn, Cd, Fe, Ni, Cu, Co, Mg, etc. Another group of ferrimagnetics is that of the **iron garnets**: YIG (yttrium iron garnet) has the formula $Y_3Fe_5O_{12}$.

The total moment of a ferrimagnetic results from a partial offsetting of the moments in one direction by antiparallel moments with a different magnitude, whereas in antiferromagnetism the resultant is zero because the moments in opposite directions have equal magnitudes. Although their maximum intensity of magnetization is reduced by the partial compensation between antiparallel moments, ferrimagnetics still have very attractive properties.

In the first place, they are *electrical insulators*, whereas usual ferromagnetics are metals. If a metal is subjected to a high frequency magnetic field, induced currents are set up in it, causing a substantial dissipation of

Table 3.2. Curie point and spontaneous magnetization at 0 K for some ferrimagnetics

Substance		Curie point (K)	Magnetic moment of the unit cell (magnetons)	Spontaneous intensity of magnetization at 0 K (10^6 A m^{-1})
Fe_3O_4		858	4.1	0.48
Spinels	$CoCr_2O_4$	98	0.18	
	$NiFeVO_4$	610	0.70	
YIG ($Y_3Fe_5O_{12}$)		560	5.00	0.13

energy as heat. The losses increase with frequency and this restricts the use of ferromagnetic metals to low frequency applications. In contrast to that, the magnetic properties of ferrites and garnets can be used with advantage even in high frequency circuits (transformers with ferrite cores). Iron garnets are magnetic materials transparent to microwaves or light and have several important applications.

Finally, it is worth pointing out that, while there are only a few ferromagnetics, a large number of ferrimagnetics have been synthesized and this has enabled a wide range of products with a great variety of magnetic properties to be created.

The behaviour of ferro- and ferrimagnetics in a magnetic field

So far, we have shown how spontaneous magnetization can occur below the Curie point because of the magnetic interaction between atomic moments, but there are many observations still to be explained. The magnetic state of a piece of iron at room temperature is extremely variable. In particular, it may be completely demagnetized: how is this consistent with spontaneous magnetization? When it is magnetized, its magnetization depends not only on the field it is currently experiencing but also on the fields to which it has previously been subjected. The behaviour of different materials with the same field variations is also highly variable.

Measurements made to establish the way strongly magnetic materials behave yield very complicated sets of observations. The theoretical idea that throws light on the results is the concept of the **Weiss domain**. Without going into details, we shall show how this concept explains, at least in outline, several of the typical properties of magnetic substances, particularly those which have important technical applications.

Weiss domains

Consider a crystal of iron at room temperature, i.e. well below the Curie point. It is observed that in a zero magnetic field the crystal may have no overall magnetic moment. Now, we know that on the atomic scale the magnetic moments of neighbouring atoms are aligned parallel to each other in the same direction. In order to resolve this apparent contradiction, Weiss put forward the idea that the crystal is divided into domains, of a size small compared with that of the whole crystal but quite large in comparison with interatomic distances (at least several micrometres). The moments are aligned within every domain, which are therefore spontaneously magnetized, but the directions of magnetization of different domains are different, so that the magnetic moment of the whole crystal can be zero (Fig. 3.9).

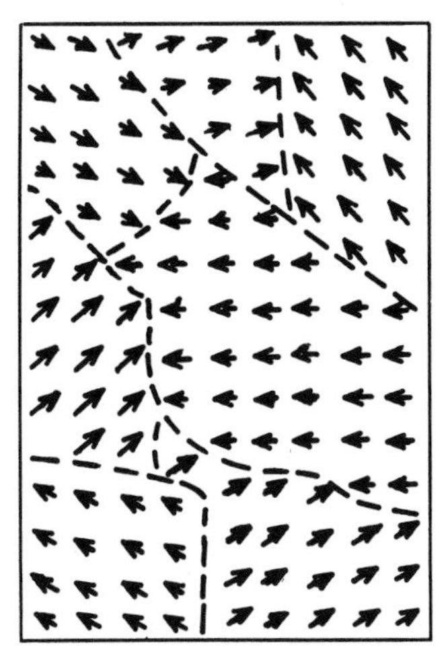

Fig. 3.9. Diagrammatic representation of Weiss domains.

The existence of such domains has now been demonstrated experimentally, and the **Bitter experiment** in particular provides a very direct demonstration of them (Fig. 3.10). The polished surface of the iron is covered with a thin layer of liquid containing a suspension of very fine magnetic particles. At the boundary between two domains with different orientations, there exists outside the metal and near its surface a magnetic field localized along the wall between the two domains. This field attracts the particles, in the same way that iron filings are attracted by the poles of a magnet, thus making the domain boundaries visible.

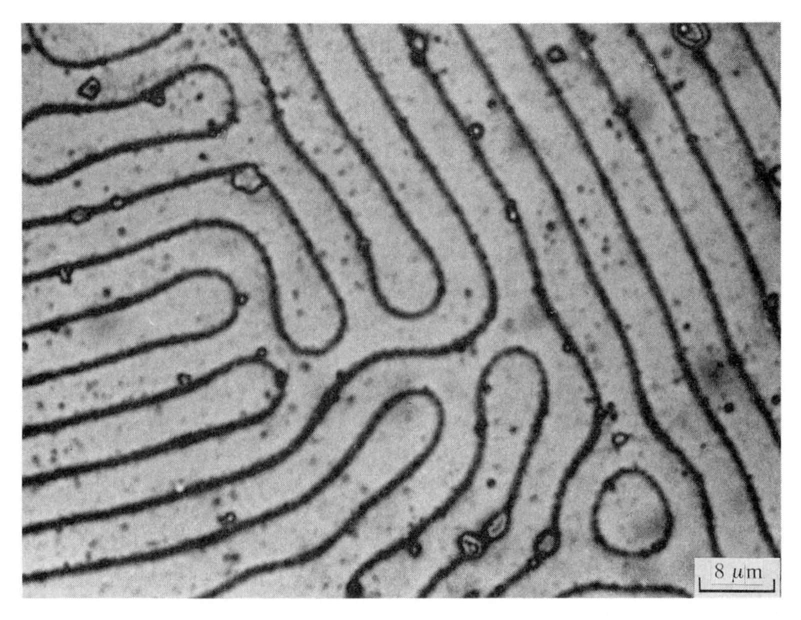

Fig. 3.10. Domains in a thin slice of garnet magnetized in both directions normal to its own plane. Domain boundaries made visible by the Bitter method (by kind permission of I. Puchalska).

The appearance of the domains varies enormously from one ferromagnetic to another. Some of the observations can be explained by the following ideas:

1. The magnetic field created by the magnetized domains outside the material has a certain energy, and the arrangement of the domains is such as to minimize this energy. That explains why, for example, domains are formed in such a way that the lines of magnetic flux, instead of emerging

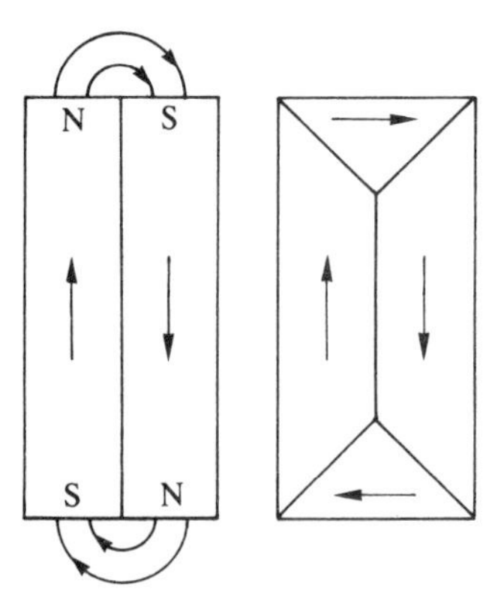

Fig. 3.11. Closure domains eliminating the external magnetic field.

into the region outside the material are turned back into the interior, thus producing no external field (Fig. 3.11).

2. The crystal is magnetized anisotropically: the atomic moments have a tendency to align themselves not just in any direction but along certain crystal axes known as directions of easy magnetization.

The hysteresis cycle

Because the magnetized state of a ferro- or ferrimagnetic depends on the history of the treatment it has had, the phenomena can only be studied quantitatively if the experimental conditions are simple and exactly specified. We shall limit ourselves to the following experimental procedure.

The specimen is inserted into a *uniform* field of *fixed* direction which is oscillating regularly between two opposite extreme values $\mu_0 H_M$ and $-\mu_0 H_M$. Under these conditions, the magnetization, for a given value of the applied field, $\mu_0 H$, has *two* values depending on whether the field $\mu_0 H$ has been reached through increasing or decreasing values. This is the hysteresis effect, characteristic of the behaviour of ferro- and ferrimagnetics. When, starting from a value of $\mu_0 H$, the field returns to the same value after having performed a complete cycle, the material has followed what is called a **hysteresis cycle** (Fig. 3.12).

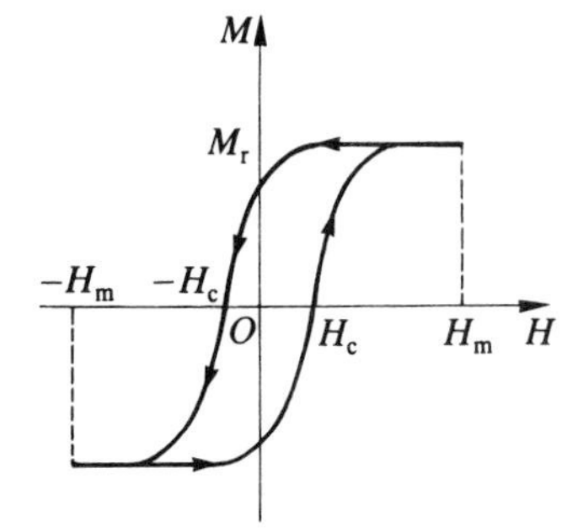

Fig. 3.12. The hysteresis cycle of a ferromagnetic.

We can explain this effect with the help of the Weiss domains. There is a particularly interesting experiment which enables the variations in magnetization and the evolution of the domains to be followed in parallel with each other (Fig. 3.13). In a thin transparent section of a ferrimagnetic garnet crystal, the domains cross the slice and matters can be arranged so that their moments are normal to the plane of the slice: they may be oriented in either direction. Examination under the polarizing microscope reveals the two types of domain (because of the magnetic rotary power of

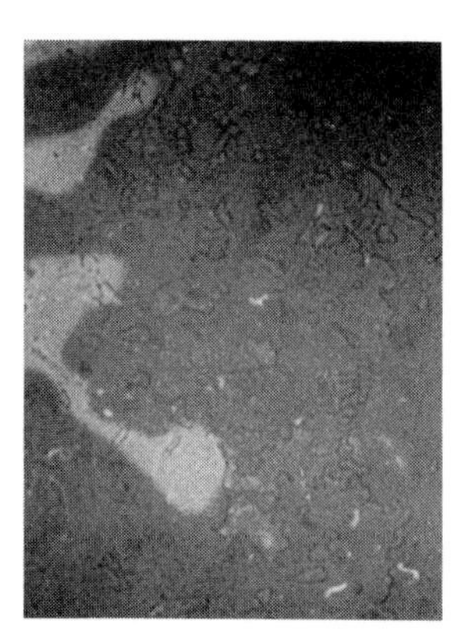
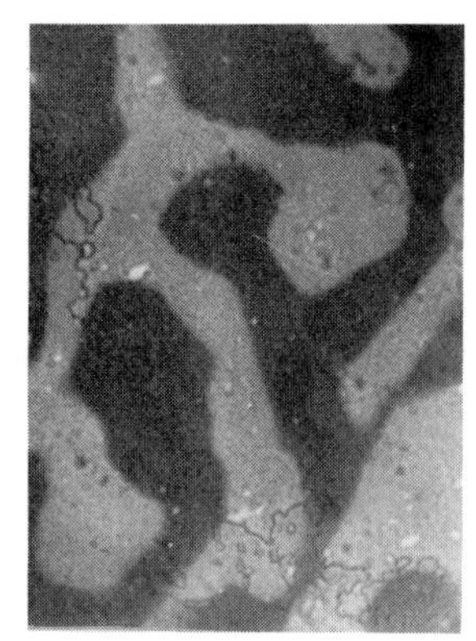
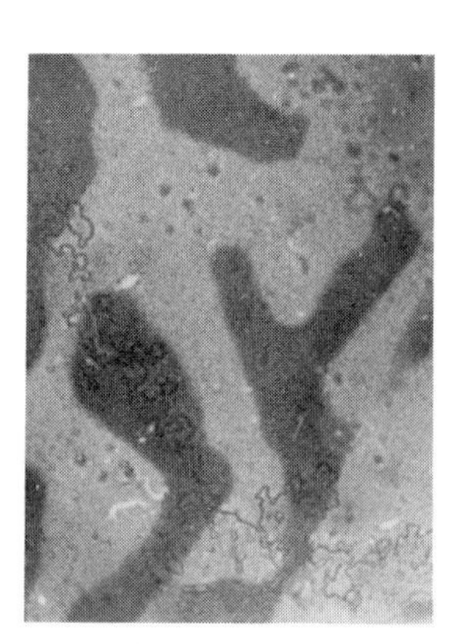

Fig. 3.13. Domains in a garnet crystal: growth of one type in an increasing magnetic field (by kind permission of M. Cagnon).

the garnet, the transmitted intensity depends on the direction of magnetization). The two types are designated, say, as positive and negative.

We start from the *saturated* state of magnetization (Fig. 3.12), where the applied field has its maximum value $\mu_0 H_M$. All the domains are 'positive', aligned in the direction of the field. The total magnetic moment is therefore simply the atomic moment multiplied by the number of atoms. The crystal has the moment already calculated for spontaneous magnetization (p. 150). If $\mu_0 H$ were increased beyond the value $\mu_0 H_M$, nothing would happen since the system is in a state of minimum energy.

Moreover, if the field is reduced slightly, nothing happens either. However, if $\mu_0 H$ continues to decrease, domains begin to appear in the negative direction: the field is no longer strong enough to oppose local fluctuations in the direction of magnetization corresponding to fluctuations in the energy of the system. When the decrease in H is continued, the overall magnetization decreases because the number and extent of the negative domains increases. When H reaches zero, the two directions of magnetization are equivalent from the energy point of view, but the positive domains were in the majority and remain preponderant. The material is still magnetized and this **remanent magnetization** or **remanence** is the 'memory' of the initial state.

The field direction is now reversed and its magnitude gradually increased. Now it is the negative domains which have the lower energy. New negative domains appear and the existing ones grow at the expense of the remaining positive domains. When the two types of domain have equal volumes, the magnetization is zero: the **coercive field**, or **coercivity**, $-\mu_0 H_c$ has been reached, the negative field necessary to make the initial positive magnetization disappear.

When $\mu_0 H$ falls below $-\mu_0 H_c$, the negative domains predominate and gradually take over the whole volume. The negative magnetization grows and reaches the negative saturation value, clearly equal in magnitude to the positive saturation value we started with. In the second half of the experiment, with $\mu_0 H$ increasing from $-\mu_0 H_M$ to $+\mu_0 H_M$, the observed effects show complete symmetry with those observed in the first half and the hysteresis cycle is complete.

The variations in magnetization arise from changes both in the orientation of the domains and in the domain volumes caused by movement of their walls. These changes can be abrupt, giving rise to what is known as the **Barkhausen effect**, demonstrated by the following experiment. A coil is wound round an iron bar in which the magnetization is changing. The abrupt alterations in the domain structure induce pulses of current in the coil which, after amplification, can be detected by a loudspeaker. The noise emissions, sharp and irregular, show that many atoms are acting together because they are linked by strong interactions. The strong interaction is precisely the reason why ferromagnetics with their high intensity of magnetization exist at all.

The form of the hysteresis cycle varies considerably from one ferro- or ferrimagnetic material to another, and the characteristics that are used to distinguish one material from another are the shape of the hysteresis loop and its three main parameters: the saturation value of magnetization, the remanence, and the coercivity. Broadly speaking, we can say that there are two extreme types between which all real magnetic materials are found and it is these two limiting cases that we now examine.

Soft magnetic materials

The hysteresis loop here is very narrow, so narrow that to a first approximation the magnetization takes the same values for the same applied field whether the field is increasing or decreasing. The single average curve can be represented as in Fig. 3.14: the magnetization M is

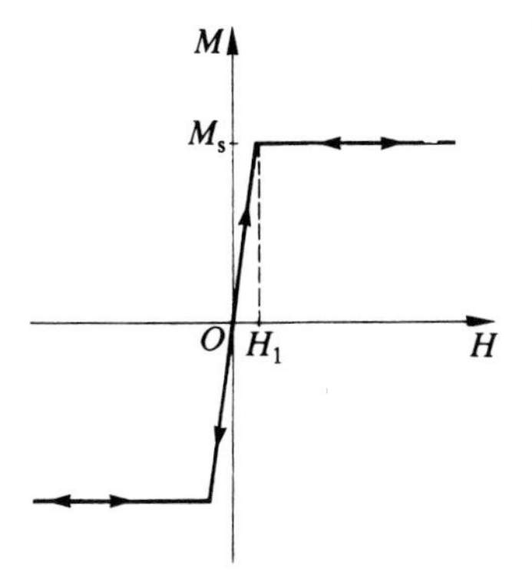

Fig. 3.14. The hysteresis cycle for a soft ferromagnetic.

proportional to the field up to saturation at the field H_1 and remains constant for $\mu_0 H > \mu_0 H_1$. When H is less than H_1, the susceptibility $\chi_m = M/H$ (or the permeability $\mu_r = 1 + \chi_m$) is constant (remember that $B = \mu_0 \mu_r H = \mu_0(H + M) = \mu_0 H(1 + \chi_m)$).

The main characteristic of 'soft' materials is a very high permeability, i.e. a high magnetic field can be obtained with a small exciting current and one which, in addition, is proportional to the current.

For pure iron, usually known as soft iron, the permeability is of the order of 1000; for silicon-iron (an alloy of iron and 3% silicon), it can be as high as 15 000 provided that the alloy is given a suitable crystal texture (SM, p. 126). Some special alloys, such as permalloy and mumetal, reach permeabilities of 10^4 to 10^5.

To bring out the significance of the high permeability, consider a current-carrying solenoid filled with a very good soft ferromagnetic in the centre of which a gap has been cut (Fig. 3.15). The field in the gap is between 1000 and 10 000 times greater than the field produced by the current alone without the ferromagnetic core as long as saturation is not reached: this greatly increased field then faithfully follows variations in the magnetizing current.

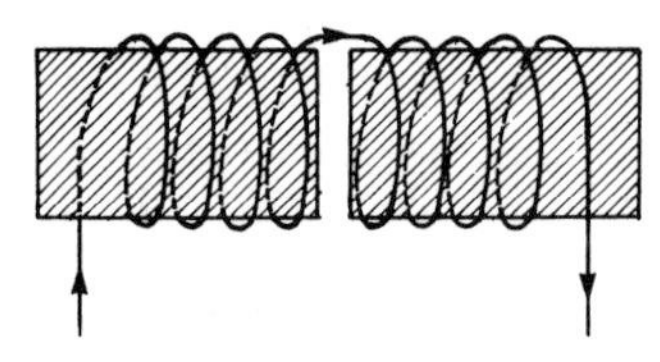

Fig. 3.15. Air gap in the iron core of a solenoid.

Very strong fields are required for the production of powerful electrical machinery. It is only because high permeability ferromagnetics have been available that we have been able to build the electric generators and motors, together with the high efficiency transformers, that are essential for the transmission of electrical energy over large distances.

There are, however, several limitations in the use of soft magnetic materials:

1. The magnetization of iron is limited by saturation. If we return to the arrangement of Fig. 3.15, the field in the gap is the sum of the field $\mu_0 H$ produced by the current in the solenoid and the field $\mu_0 M$ created by the magnetized iron. The latter is limited to its saturation value of 1 to 2T depending on the material. When the field in the gap is less than 1T, the effect of the iron is overwhelmingly predominant. For very strong fields, however, the field due to the current becomes the greater of the two (Fig. 3.16), so that iron-cored coils have no advantage over air-cored coils and we are reduced to producing intense fields by using very large currents.

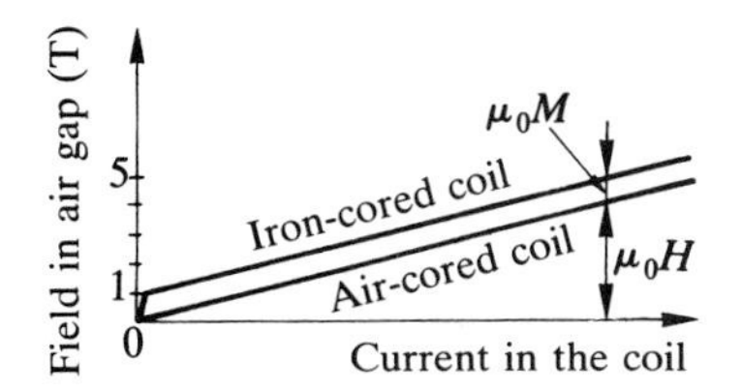

Fig. 3.16. Magnetic field in the air gap of Fig. 3.15 as a function of the magnetizing current. The increase in the field due to the presence of the iron is considerable for fields below 1 T, but very small for a field of 5 T.

This is technically very difficult: either a considerable amount of energy is wasted in the resistance of the conductors, which has to be dissipated by rapid circulation of cooling water, or we have to use superconducting circuits with their need for the production of low temperatures. It is easy to see why new superconductors with high critical temperatures are awaited with so much impatience.

2. Soft magnetic materials are generally used in alternating fields and, although the hysteresis loop is very narrow, it does have a finite area which corresponds to the dissipation of heat in the material per cycle. The wasted power is proportional to the area of the loop and the frequency of the alternating field, so that it becomes large at high frequencies. For that reason, attempts have been made with some success to produce soft magnetic materials with very low magnetic losses.

3. There is also, of course, a second cause of energy wastage: electrical losses due to electromagnetically induced currents in cores and armatures, which can be reduced in metals by using stacks of thin sheets separated by insulating layers. These losses can also be prevented by using an insulating material for the cores, and this has led to the use of ferrites (p. 159), whose discovery greatly facilitated the production of high frequency electric and electronic circuits.

Hard magnetic materials

These are characterized by a very wide hysteresis loop, i.e. by high remanence and coercivity. An extreme case is that of a rectangular cycle: the magnetization remains almost at its saturation value up to a reverse field of $-\mu_0 H_c$, and it then falls abruptly to the opposite value for any field $\mu_0 H$ below $-\mu_0 H_c$ (Fig. 3.17).

Whereas soft materials are used to reinforce the magnetic field of a current and faithfully follow its variations, hard materials are used for the production of **permanent magnets**. The material is magnetized to saturation and the field is removed: the material then remains magnetized. It would seem that the magnetization should be equal to the remanence since the applied field is zero. In fact, the situation is not quite so simple: the block of magnetized material produces its own magnetic field called the

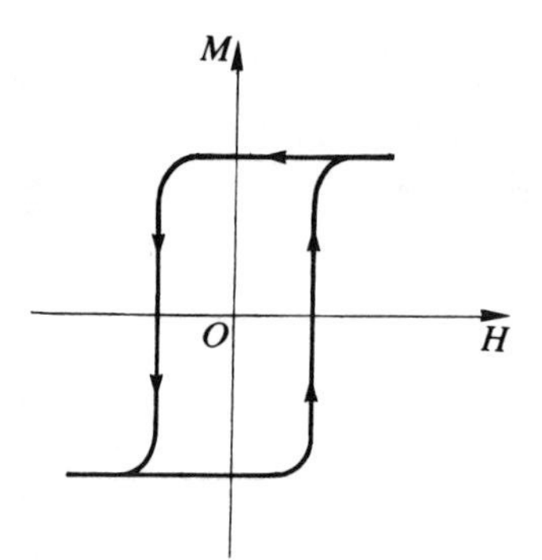

Fig. 3.17. A 'rectangular' hysteresis cycle for a hard ferromagnetic.

demagnetizing field, which depends on the external shape of the block. It varies throughout the volume of the magnet, but is always in the opposite direction to the magnetization, which is therefore reduced to a value below the remanence. Not only that, but during use the magnet may find itself in magnetic fields from various external sources and such fields might be in the opposite direction to the magnetization. If the coercivity is low, the magnet could then be accidentally demagnetized.

If the 'permanent' magnetization is to be strong and stable, therefore, both the remanence and the coercivity should be high. More precisely, a 'quality factor' is used for permanent magnet materials consisting of the maximum value of the product $\mu_0|HM|$ over that part of the hysteresis loop between the points ($M = M_r$, $H = 0$) and ($M = 0$, $H = -H_c$) (Fig. 3.12). With older magnet steels, this parameter was $250\,\mathrm{J\,m^{-3}}$. Much better ferromagnetic alloys are now known, such as alnico ($2000\,\mathrm{J\,m^{-3}}$) and ticonal ($8000\,\mathrm{J\,m^{-3}}$). A ferrite known as ferroxdur has a quality factor of 1200 and a coercivity of 0.2 T, where the latter was only 0.05 T in the older magnet steels.

Advances in hard magnetic materials have enabled the fields produced to be increased and, at the same time, their volumes to be considerably reduced. With the older magnet steels, long bar magnets or large horseshoe magnets were necessary to produce sufficient reduction in the demagnetizing field. Today, using ferrites, small 1 cm cubes are strong magnets (used for door closure, electrical measuring equipment). The repulsive force between the two like poles of two ferrite magnets can be used to make one float above the other (Fig. 3.18), since it is sufficient to counterbalance the weight.

Ferrites like (CoFe)O, Fe_2O_3 have highly rectangular hysteresis loops (Fig. 3.17) and were used as memories in first generation computers. To a good approximation, ferrites can be said to have only two possible states of magnetization, which can be denoted by 0 and 1. In magnetizing the ferrite by a field stronger than the coercivity, a 0 or 1 is 'written' into it, depending

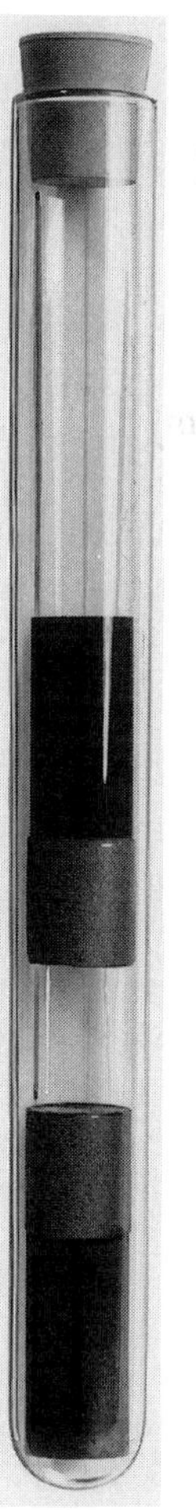

Fig. 3.18. A magnet levitated by the repulsive force between two like poles. (Courtesy of Palais de la Découverte.)

on the direction of the magnetizing current. This figure is kept in the memory as long as no field is applied with a magnitude greater than $\mu_0 H_c$. The figure stored in the memory can be read without erasing it by using a signal induced in another circuit.

Relationships between magnetic properties and the structure of matter

Can the great variety in magnetic properties be explained by the atomic structure of matter? In fact, we are not in a position to account for every peculiarity that might be observed and above all we cannot predict structures that will exhibit a particular macroscopic property. Nevertheless, considerable progress has been made in the theories of magnetism since the 1930s, in close harmony with the increase in empirical knowledge. We shall merely give one example here: we describe the general idea that enables us to understand the existence of soft and hard magnetic materials.

In a soft material, the magnetization faithfully follows variations in the applied field. The volumes of the domains therefore change almost immediately and this means that the domain walls move through the crystal without encountering any resistance. It is easy to see that this will be easier in a lattice without irregularities and in a crystal without impurities. In fact, soft iron is a high purity iron. Moreover, it is observed that, in high permeability materials, any mechanical strain, however slight, reduces the permeability considerably.

On the other hand, when the applied field is varied in hard magnetic materials, the domains have a tendency to remain as they are, as least as long as the gain in energy by reversing the domain is insufficient. This is the type of behaviour that is almost exactly like that observed in ferrimagnetics with rectangular hysteresis loops. Thus, in hard magnetic materials, the domain walls only move with difficulty, and they are **pinned** at anchoring points from which they can only be freed at the cost of a certain amount of energy. These anchoring points may be small grains of a second phase or lattice defects. That is why hard materials are always observed to have complex structures and are often alloys with a complicated composition.

4

Mechanical properties of solids

General survey

So far, we have been examining properties of solids which lead to specialized applications for certain classes of them, e.g. as electrical conductors or magnets. However, for the vast majority of solids surrounding us, it is only their **mechanical properties** that are of any concern, i.e. their behaviour under the action of the forces to which they are subjected.

As a result, we could say that this is a question of 'non-properties' rather than properties. After all, the normally accepted and most elementary meaning attached to the word 'solid' is that of resistance to various forms of onslaught without apparent change: resistance to pressure, tensile stress, twisting, vibration, collision, friction, etc, as well as to attack by atmospheric agents: water, air, and so on. Solid building materials, for example, have been chosen from the earliest times for their ability to provide a long-lasting and reliable shelter from the elements. Weapons and tools from the very beginning, and machines in later periods, have been made in such a way that, as a primary objective, they did not break, deform, or wear away.

However, there is a limit to the unalterability of any solid: when the external action exceeds a certain threshold, permanent deformation or breakage occurs. The behaviour of solids varies widely, both qualitatively (e.g. some are brittle, some ductile) and quantitatively (e.g. in their ultimate strength, durability, and ability to withstand external forces). Thus, glass, which is brittle, breaks without undergoing any deformation. Metals, which are ductile, deform without breaking and among them some are rather soft and others very hard. This has led human beings, from the very beginning of their history, to seek out and select materials for their specific properties: flint for cutting tools, wood because it can easily be shaped, steel for weapons, bronze for implements, etc.

When these artefacts were first produced, the conditions under which they were used subjected the materials in them to stresses that were far below the limits beyond which they would fail, so that an accurate knowledge of such limits and of the behaviour of the materials in extreme conditions were not necessary: the Romans building an aqueduct did not worry about the exact value of the maximum load that could be supported by the blocks of stone without damage. It was technological improvements that led to the need for a better use of the properties of materials. The safety factor (the ratio of the maximum stress that a material can support to the stress it is subjected to in normal use) is still enormous in present-day houses, but is reduced for an aircraft.

As a consequence of this, more extensive and detailed studies of the behaviour of solids under stress have become increasingly necessary. It is because of such studies that safety factors can be deliberately lowered, which is often a precondition for the development of certain projects: e.g. because of the need to reduce the production cost of machinery or because an aircraft can only be built if there is sufficient knowledge to produce a reliable structure that is both light and strong.

Thus, the concept of an unalterable solid is only valid over a certain restricted range whose limits must be ascertained, and the behaviour of the solid beyond these limits must then be determined.

Moreover, it appears that, even within its range of validity, the idea of unalterability is only a first approximation: the solid is in fact deformed even by very small forces, but only by a very small amount and even then its original state is recovered when the external stress is removed. A detailed knowledge of this behaviour, known as **elasticity**, is needed by engineers if they are to design a machine correctly.

In spite of the initial, somewhat superficial, impression, the mechanical properties of solids thus constitute a complex body of knowledge which is of considerable importance to many techniques. Engineers must have at their disposal a complete catalogue of all the various materials available in order to be able to choose the best for their purposes and to use them under optimum conditions. A catalogue of that type is the result of purely empirical tests and is therefore not something that concerns us in this book.

Our purpose is to relate the mechanical properties to atomic structure and to the elementary interactions between atoms. The reader might well wonder why, given the role of mechanical properties in everyday life, we did not begin the book with a study of this aspect of solids. The reason is that we are still, and are likely to remain for a long time to come, a long way from our objective, i.e. from an understanding of, and an ability to predict, the mechanical behaviour of solids from the properties and geometrical arrangement of the atoms contained in them.

The models which proved to be so useful in studying the thermal, electrical, and magnetic properties are fundamentally inadequate for the mechanical properties, and those that would be valid are so complex that they cannot be completely exploited to give quantitative results.

We shall see that an essential role is played by rare and very small

structural defects which interrupt the regular arrangement of the atoms in an ideal crystal. These are the entities we shall be dealing with throughout the chapter: the 'crystal defect' will be a sort of leitmotif constantly recurring in our account. Yet we cannot, unfortunately, obtain an accurate knowledge of the defects existing in a given specimen.

We find ourselves here at the centre of major difficulties in the physics of solids and are forced to recognize that some of them are in practice insurmountable. Nevertheless, it would not be right for us to make this introduction to the study of mechanical properties too pessimistic, for it must also be said that the physics of solids has had an important role to play in the advances made over the last fifty years. The dense and tangled thicket of raw experimental data has been unravelled, clarified, and generally understood in a qualitative fashion. Although physicists cannot use first principles to achieve any numerical result needed by technologists, the general ideas they have developed have guided experimental research and without them it would not have been possible to create the new materials which are the basis of the advances made in major technologies.

Elasticity

We consider one of the simplest possible examples of an elastic system: a metal wire attached at its upper end to a fixed support and with a load of variable mass suspended from its lower end. The distance between two marks engraved on the wire is measured very accurately as the load is increased. It is observed that the wire becomes longer and the following facts are noted:

1. The extension is proportional to the load.

2. For all solids, apart from some exceptional materials, the relative extension is very small: e.g. for mild steel, a length of 10 cm increases by 0.05 mm under a stress of 100 MPa.[†]

3. If the load is removed, the wire recovers its original length exactly. However, this is only true if the stress is less than a certain limit, e.g. approximately 200 MPa for mild steel.

These are the characteristic features of an elastic deformation or strain. Some aspects are worth examining in more detail.

Normal solids, i.e. those in which the elastic strain is small, include all crystalline materials whatever the type of interatomic bond, together with glasses or amorphous bodies. For strong solids, the maximum elastic strain

[†]A stress (force per unit area) of 10 megapascals (MPa) is, to a very good approximation, equal to the weight of a mass of 1 kg acting over an area of 1 mm^2.

is less than 1 per cent and the stresses they can withstand without damage are considerable: several GPa. That is why it is a good approximation to regard solids as unalterable under normal conditions.

There are, however, exceptional solids with a very high elasticity, a typical example being rubber. A strip of rubber can be extended by 100 per cent under a stress of only 1 MPa. The strip recovers its original length when the stress is removed. Rubber, and all the materials with similar elastic properties, are formed from long chain polymers with a fairly disordered arrangement. Such structures are described in texts dealing with polymers (e.g. SM, p. 187) and they show how the structure explains the unusual elastic properties. Whereas a metal under stress retains the same atomic structure with very little change in its crystal parameters, the structure of rubber does change: the polymer chains, curled up in the unstressed state, unwind under stress. In the rest of our treatment of solids, we shall only consider elasticity of the type exhibited by metals.

A very common piece of apparatus, based on the elasticity of metals, is the coiled spring (often used in elementary physics courses). We know that the spring extends considerably with quite small forces, but this does not contradict what we have been saying previously about the elastic deformation of metals. This is because the change in the shape of the coil is the result of the movement of different segments of the wire under tension. But each segment, in itself, has suffered only a small strain, as is shown by an analysis of the extension of such a spring.

Box 17

Geometry of the extended spring

A spring is a metal component so designed that a considerable overall extension can be obtained with only very small local strains. Thus, while each small section of the material remains elastically deformed, a linear relationship between force and displacement is obtained over an easily observable range of extensions.

Take the example of a coiled spring (Fig. 4.1), formed from a wire wound into a helix with a circular cross-section of radius R. The total length of the wire is denoted by l and this is clearly much longer than the length of the spring, h. The helix has n turns.

With the upper end A attached to a rigid support, a force F is applied to the lower end B in a vertical direction as shown and the turns of the coil separate from each other, producing a total extension of Δh in the spring as a whole. Consider a half-turn M_1MM_2 in isolation, and replace it by a flexible joint M between two rigid pieces

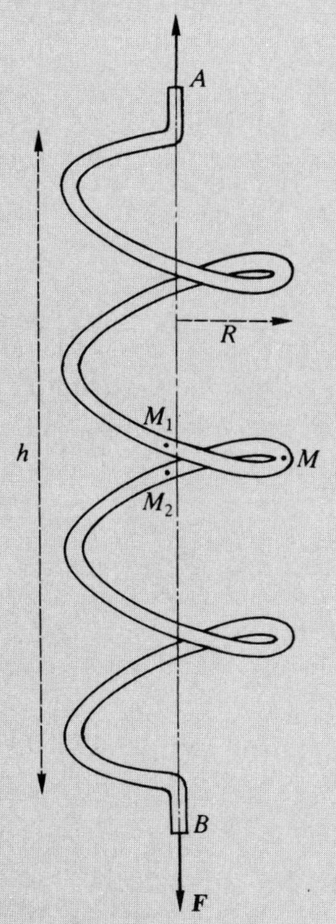

Fig. 4.1. Elastic extension of a coiled spring.

of wire M_1M and MM_2. The joint M is subjected to a torque of moment FR and, since the vertical distance between M_1 and M_2 has increased by $\frac{1}{2}\Delta h/n$, the joint experiences a twist of:

$$\Delta\alpha = \frac{1}{R}\left(\frac{1}{2}\frac{\Delta h}{n}\right).$$

In reality, the twist is not localized at M, but is distributed along the wire. The angle of twist of the wire per unit length is:

$$\frac{\Delta\alpha}{\dfrac{l}{2n}} = \frac{\Delta h}{lR}.$$

The intrinsic strain of the wire is that of a wire of length l twisted through an angle $\Delta\theta = \Delta h/R$ under the action of a torque of moment FR. If C is the torsional constant of the wire, we have that:

$$FR = C\,\Delta\theta = C\,\frac{\Delta h}{R}\,.$$

This gives the following relationship between F and Δh:

$$F = C\,\frac{\Delta h}{R^2}\,.$$

The theory of elasticity also enables us to calculate the constant C, which is related to the rigidity modulus μ (defined later in Box 18). The expression for C is:

$$C = \frac{\pi}{32}\,\mu\,\frac{d^4}{l}$$

where d is the diameter of the wire.

What happens when the stress on a solid exceeds the elastic limit? The behaviour displayed by different materials covers a very wide range but, for the moment, we merely distinguish two extreme types since many materials have properties intermediate between them. The first type is exhibited when the specimen used in our initial experiment remains stretched when the stress is removed, leaving a permanent strain. If the elastic limit is greatly exceeded, the permanent strain can be of the order of several per cent, i.e. much greater than the maximum elastic strain. The external shape of the solid is therefore changed considerably, yet its internal cohesion is maintained. This behaviour is known as **plastic deformation** or **ductility**: most metals are ductile.

The other possibility is that the specimen, without an appreciable change of shape, suddenly breaks into two. Such a material is said to be **brittle**, and this is typical of ionic and covalent crystals and glasses.

In this chapter, we shall be examining both ductility and brittle fracture, together with the relationship between them. First of all, however, we look at elastic behaviour, i.e. the state in which any deformation is completely reversible. The elastic and plastic regions are in fact continuous, and the elastic limit cannot be precisely defined since it is difficult to determine the onset of a permanent strain: the result depends on the sensitivity of length measurements. However, this is not of fundamental significance and we shall not discuss measuring techniques.

The theory of elasticity in mechanics

We first considered a very simple case: that of a solid in the shape of a cylinder subjected to a single force along its axis. We now take a solid, assumed to be homogeneous, of any shape and subjected to a general system of forces. Provided the stresses at any point in the solid do not exceed the elastic limit, the theory of elasticity enables the strain at each point to be calculated from the system of forces (or vice versa). The behaviour of the material is defined by a certain number of coefficients, the elastic moduli, which are measurable. Mechanical engineers are not concerned with the structure of matter on the atomic scale, nor with the properties of the constituent atoms: for them, matter is completely modelled by the macroscopic empirical coefficients and they do not question their origin.

A homogeneous solid may be anisotropic: the strain produced by a stress then depends not only on the magnitude of the stress but on its direction as well. This is what generally happens in crystals. The number of coefficients characterizing the solid varies with the lattice symmetry. The calculations are more complex in the anisotropic case, but the theory copes perfectly well with them (it was precisely for this purpose that mathematicians developed tensor calculus).

To give some idea of the bases of the theory of elasticity, we shall stick to the case of an **isotropic** solid. This can still embrace the case of a crystalline material since, if the solid is formed from a large number of very small crystal grains stuck to each other, the whole system is isotropic even though each grain might be anisotropic, provided that the axes of different crystallites are randomly oriented in space.

Consider a small cube of material with edges (equal) of l_1, l_2, l_3. It is subjected to a stress σ (force per unit area) parallel to one edge, l_1 for example (Fig. 4.2). The cube is extended by Δl_1, which is proportional to σ

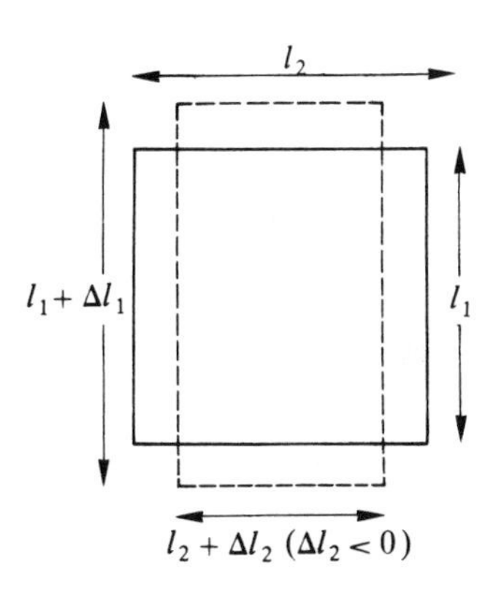

Fig. 4.2. Elastic deformation of a uniform cube.

since we are in the elastic region. The coefficient of proportionality between the stress σ and the strain, $\Delta l_1/l_1$, is known as **Young's modulus** E, so that:

$$\Delta l_1/l_1 = \sigma/E.$$

At the same time as it is extended, the small cube also contracts laterally in such a way that the changes in l_2 and l_3 satisfy:

$$\frac{\Delta l_2}{l_2} = \frac{\Delta l_3}{l_3} = -\nu \frac{\Delta l_1}{l_1}$$

where ν is **Poisson's ratio.**

Poisson's ratio and Young's modulus are the two parameters which are sufficient to define the elastic properties of a given material if it is isotropic. It should be pointed out that, if the elastic strain of the specimen brought about no change in volume, the value of ν would be $\frac{1}{2}$. In fact, it is always less than $\frac{1}{2}$, indicating that the volume always increases (the relative increase is very slight, of the order of 1/1000 for a 0.3 per cent strain).

Young's modulus can be measured by the extension test described at the beginning of this chapter. It can also be derived from a measurement of the speed of sound in the solid. Poisson's ratio is determined by a second test, as we shall see from Box 18. This box deals with two examples of the calculation of elastic strain from E and ν: first, the **compressibility** or change of volume under a hydrostatic pressure (uniform in all directions); secondly, the **shear**, or change of shape of the cube under a stress parallel to one of its faces.

Box 18

Expressions for the compressibility and rigidity in terms of Young's modulus and Poisson's ratio

Consider a parallelepiped with its edges parallel to the x, y, and z axes of a right-handed three-dimensional orthogonal coordinate system, and of length l_1, l_2, l_3. Subject its faces in pairs to forces F_1, F_2, F_3 normal to the surfaces (Fig. 4.3) and thus to positive stresses:

$$\sigma_1 = \frac{F_1}{l_2 l_3}, \sigma_2 = \frac{F_2}{l_3 l_1}, \sigma_3 = \frac{F_3}{l_1 l_2}.$$

Under the action of these forces, the edges will extend by Δl_1, Δl_2,

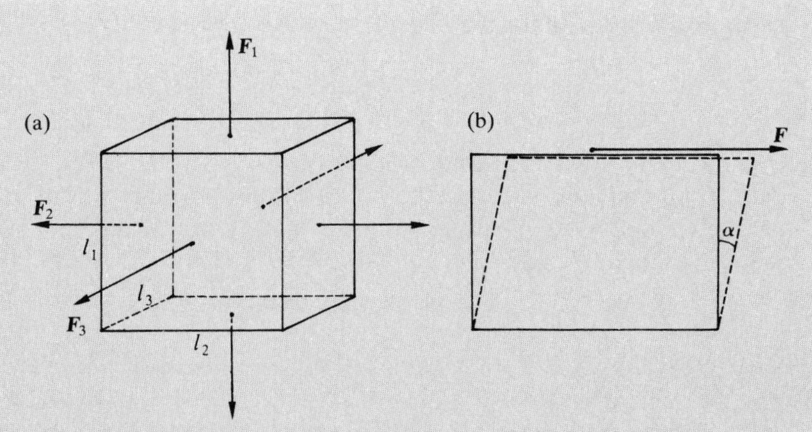

Fig. 4.3. (a) Deformation under an isotropic stress; (b) deformation under a shear stress.

Δl_3. Denote the relative extensions (i.e. the tensile strains) by e_1, e_2, e_3:

$$e_1 = \frac{\Delta l_1}{l_1}, e_2 = \frac{\Delta l_2}{l_2}, e_3 = \frac{\Delta l_3}{l_3}.$$

The theory of elasticity postulates that there is a linear relationship between the stresses and strains. In the experiment described to define E and ν, we applied a single stress σ_1, the two others being zero ($\sigma_2 = \sigma_3 = 0$). From the definitions of E and ν, we then had $e_1 = \sigma_1/E$ and $e_2 = e_3 = -\nu e_1$. By imagining that the stresses σ_1, σ_2, and σ_3 are applied in turn to the three faces and by adding the respective strains algebraically, we obtain:

$$e_1 = \frac{1}{E}[\sigma_1 - \nu(\sigma_2 + \sigma_3)], e_2 = \frac{1}{E}[\sigma_2 - \nu(\sigma_3 + \sigma_1)],$$

$$e_3 = \frac{1}{E}[\sigma_3 - \nu(\sigma_1 + \sigma_2)].$$

We now show that these relationships enable us to find expressions for the two coefficients, the **compressibility** and **rigidity modulus**, involved in two very common types of stress.

If a solid is subjected to a uniform pressure over the whole of its external surface, the relative change in volume (negative) $\Delta V/V$ is given as a function of the pressure ΔP by the **compressibility** K:

$$K = -\frac{1}{V}\frac{\Delta V}{\Delta P}.$$

(Note that in some texts, the reciprocal of the compressibility is defined instead: this is called the **bulk modulus**.) In the case of our parallelepiped, we see that the application of a pressure ΔP amounts to the application of equal negative stresses $\sigma_1, \sigma_2, \sigma_3$:

$$\sigma_1 = \sigma_2 = \sigma_3 = -\Delta P$$

over the three faces. In addition, the relative volume change can be expressed simply in terms of the extensions:

$$\frac{\Delta V}{V} = \frac{\Delta(l_1 l_2 l_3)}{l_1 l_2 l_3} = \frac{\Delta l_1}{l_1} + \frac{\Delta l_2}{l_2} + \frac{\Delta l_3}{l_3} = e_1 + e_2 + e_3.$$

Using the expressions for e_1, e_2, and e_3 obtained above, we then obtain:

$$\frac{\Delta V}{V} = e_1 + e_2 + e_3 = \frac{1-2\nu}{E}(\sigma_1 + \sigma_2 + \sigma_3) = -\frac{3(1-2\nu)}{E}\Delta P$$

and hence:

$$K = \frac{3(1-2\nu)}{E}.$$

A shear stress involves the application of a force on the upper surface of the parallelepiped parallel to that surface, while keeping the lower surface fixed (Fig. 4.3).

It can be shown that the angle of shear α of the lateral surfaces depends only on the ratio F/S, where S is the area over which the tangential force creating the shear stress is applied. Below the elastic limit, α is proportional to F/S, and the **rigidity modulus** μ is defined by:

$$\alpha = \frac{1}{\mu}\frac{F}{S}.$$

We now show that μ can also be expressed in terms of E and ν. Consider a cube (Fig. 4.4) of edge l, and let stresses σ_1 and σ_2 be applied normally to the surfaces perpendicular to the plane of the

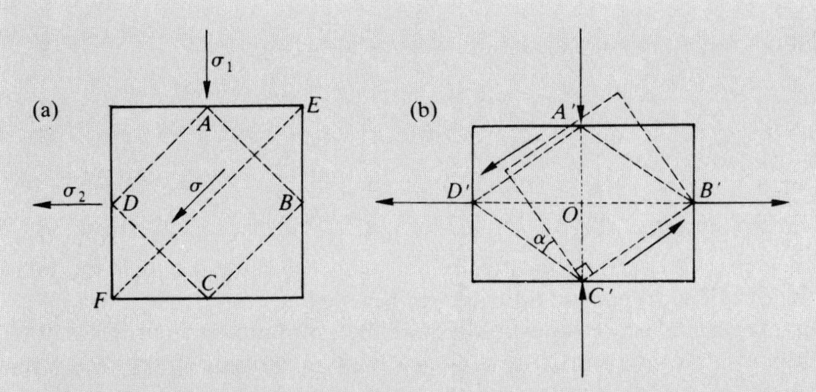

Fig. 4.4. Calculation of the rigidity modulus from Young's modulus and Poisson's ratio: (a) cube before deformation; (b) cube after deformation (exaggerated for clarity).

figure in such a way that the expansion of the whole volume

$$\frac{\Delta V}{V} = e_1 + e_2 + e_3$$

is zero. Since σ_3 is zero, one of the previous equations gives us that $\sigma_1 + \sigma_2 = 0$, and hence that $\sigma_2 = -\sigma_1$. We then have that the extensions in the two directions are equal and opposite:

$$e_1 = -e_2 = \frac{1+\nu}{E}\sigma_1 .$$

The pair of stresses σ_1 and σ_2 produces a shear along the diagonal plane denoted by *EF*: the square *ABCD* is changed into a rhombus $A'B'C'D'$. To the first order, $A'D'$ is equal to *AD* and the distance between $A'D'$ and $B'C'$ is still equal to *AB*. The angle of shear α is such that $\alpha + \pi/2 = \angle D'C'B'$. Now:

$$\tan\frac{\alpha + \frac{\pi}{2}}{2} = \frac{OD'}{OC'} = \frac{1+e_1}{1-e_1}.$$

Since the elastic strains are always small, this is equivalent to $\alpha/2 = e_1$.

Moreover, the resultant of the applied forces $\sigma_1 l^2$ and $\sigma_2 l^2$ is $\sqrt{2}\,\sigma_1 l^2$. However, this force is exerted over an area of $\sqrt{2} l^2$. The shear stress is thus simply σ_1. Using the definition of μ and the relationships between e_1 and σ_1, we find:

$$\mu = \frac{E}{2(1 + \nu)}.$$

If a compressive axial stress is applied to the specimen instead of a tensile stress, it contracts elastically and with the same Young's modulus, but the elastic limit may be different for tensile and compressive strains.

Data for the elastic properties of several common metals are assembled in Table 4.1 to give some idea of orders of magnitude. This shows that materials are available with a wide range of properties.

The elastic properties of some real materials

So far, we have described only the simplest types of elasticity, but there are materials in which the strain, while being completely reversible, is not proportional to the stress over the whole elastic region. This is the situation in elastomers like rubber, where Young's modulus decreases with increasing strain.

We might also mention the case of wood, because of its practical importance. It has the special feature that it deforms differently under compression and tension (Fig. 4.5). This is because it is a composite material, with elastic fibres incorporated in a rather soft matrix of lignin. The interactions between the two constituents are not the same under tension and compression.

The elastic moduli are determined from systematic tests carried out under simple and well-defined conditions. At present, the mechanical theory of elasticity enables the behaviour of a material to be predicted in

Table 4.1. Elastic constants of some common metals in polycrystalline form

Metal	Young's modulus (10MPa)	Poisson's ratio	Elastic limit (10MPa)	Melting point (K)
Aluminium	6900	0.33	1.22	933
Copper	12800	0.36	34	1356
Lead	1400	0.4	0.6	600
Tungsten	34000	0.28	10	3653
Iron	12000	0.17	26	1809

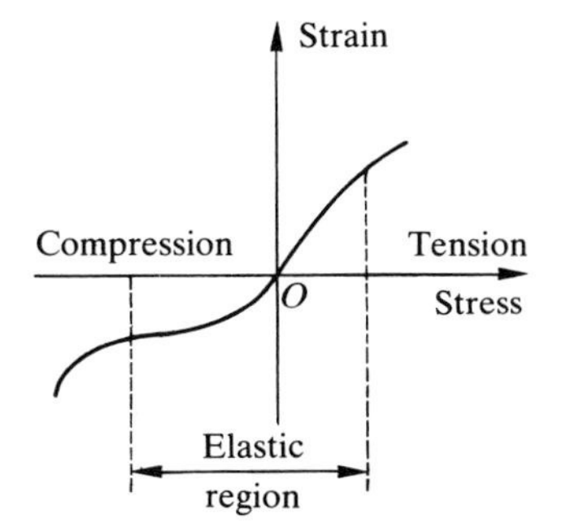

Fig. 4.5. The stress–strain relationship for wood in the elastic region.

any situation provided it remains in the elastic region: the needs of the engineering design and manufacturing sector can thus be satisfied.

On the other hand, the theory cannot answer the questions which are the main concern of this book: what are the relationships between the atomic structure of a solid and its elastic properties and elastic limit? Is it possible, for example, to calculate the elastic moduli from properties of atoms and their mutual arrangement?

Elasticity and atomic structure

Consider a crystal whose elasticity is more or less isotropic. We thus leave aside for the moment the very serious difficulties arising from the fact that adjacent grains in a polycrystalline solid are differently oriented and may be deformed in different ways.

The macroscopic elastic strain is reflected by a similar strain at the scale of the crystal unit cell: the atoms therefore move through relatively small distances and the atomic structure is only slightly modified.

Consider a cubic crystal with a unit cell of side a subjected to a tensile stress parallel to a cubic axis. Transposing the results obtained on the macroscopic scale, it can be predicted that the cubic unit cell changes into a tetragonal cell with parameters $a(1 - \nu e)$, $a(1 - \nu e)$, $a(1 + e)$. This deformation is confirmed by experiment: each family of lattice planes in the cubic crystal is transformed, in the stressed crystal, into a family of reticular planes which is very similar but whose separation is slightly different, depending on the orientation of the planes with respect to the stress. Thus, the interplanar spacing changes from a to $a(1 + e)$ for planes normal to the stress axis and from a to $a(1 - \nu e)$ for planes parallel to the other two cube faces. There are X-ray diffraction methods sensitive enough to measure these small variations in interplanar spacings very accurately, and the measurements confirm the theory. It is also possible, using the measurements of the elastic strains in the unit cell, to deduce the system of stresses to which the metal is subjected from the values of E and

ν. This provides a *non-destructive* method of finding the internal stresses in a material, which are important data for an engineer.

We must now address the main question: is it possible to calculate the coefficients E and ν for a given crystal, if we retain the approximation that the solid is isotropic?

The temperature is taken to be 0 K so that we can neglect atomic vibrations. With the crystal in a free state, the interatomic forces balance each other and equilibrium is attained when the atoms occupy the sites described by the crystal structure. Put another way, we can say that, in the equilibrium state, the potential energy of the whole crystal assembly is a minimum.

Consider first the simplified case of a pair of atoms, one we have already looked at (p. 44). There is both an attractive and a repulsive force between these atoms, the latter arising because their electron clouds cannot penetrate each other. The pair thus possesses a potential energy, whose derivative with respect to the distance separating them gives the sum of the opposing forces.

It is easy to imagine the shape of the curves giving the potential and force as a function of distance apart (Fig. 4.6). At the equilibrium distance a_0, the energy is a minimum. For $a < a_0$, the resultant force is repulsive and for $a > a_0$ it is attractive. The minimum value of the potential energy is the work needed to separate the atoms from a_0 to infinity. For a very small change in the separation around the value a_0, the displacement is proportional to the force exerted and the atoms return to their equilibrium positions when the force is removed: this is the elastic region.

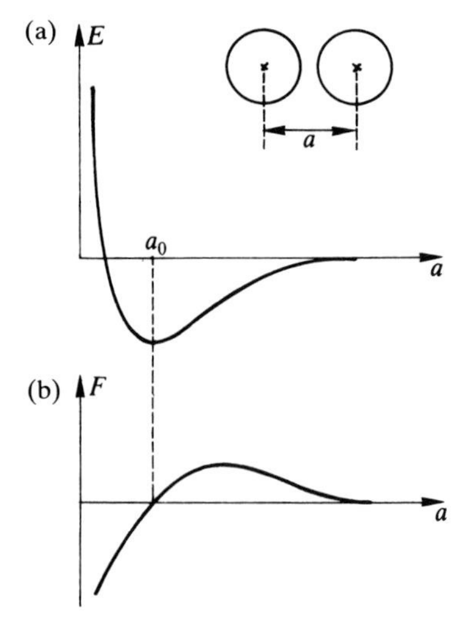

Fig. 4.6. (a) The potential energy between a pair of atoms and its variation with distance apart; (b) the corresponding force between the atoms.

Various mathematical functions have been suggested empirically to represent an energy curve similar to that of Fig. 4.6. On the other hand, and this is an important fact, we are quite unable to determine precisely the potential energy of a pair of atoms from a knowledge of their configuration in the free state, i.e. from the nature of the nucleus and the wave function describing the electron cloud around it. If we cannot solve the case of an isolated pair, it follows that we cannot do so for the crystal, which is more complex since each atom has many neighbours.

The close-packed hard sphere model is unable to represent the elastic deformation of the unit cell. To see that this is so, take a copper crystal (face-centred cubic) subjected to a tensile stress parallel to a cube edge. Consider the group of six atoms at the vertices of an octahedron inscribed in the unit cell (Fig. 4.7): nearest neighbours are 0.256 nm apart, and atoms (1) and (2) at opposite vertices are 0.362 nm apart. For a 1 per cent extension, the separation of atoms (1) and (2) changes by 4 pm. It can be appreciated how small this is by comparing it with the mean amplitude of the atomic vibrations in copper at room temperature, which is five times greater. Considering only the four atoms (1) to (4): when (1) and (2) separate, atoms (3) and (4), in order to remain in contact with the first two, will approach each other, which would explain the lateral contraction accompanying the axial extension. But this is impossible because of the presence of atoms (5) and (6), if it is assumed that the atoms in contact are hard spheres. We are therefore led next to a model in which the atoms are slightly deformable. When the tensile stress changes the equilibrium state of the free crystal, the atoms are not only displaced but are deformed as

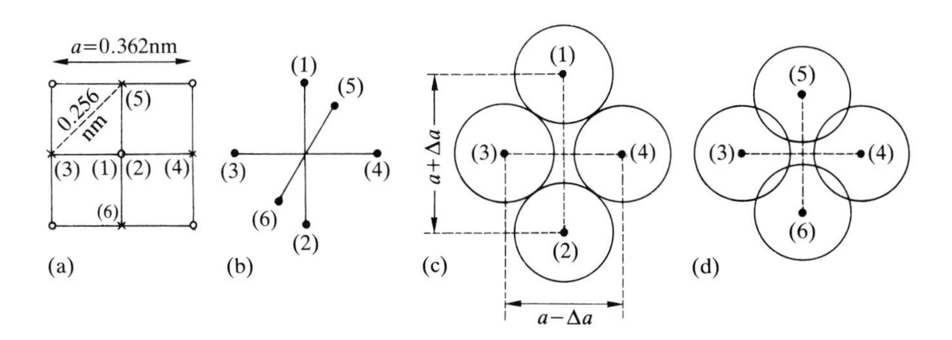

Fig. 4.7. (a) The face-centred cubic unit cell of copper; (b) the positions of the six atoms at the vertices of the octahedron inscribed in the unit cell; (c) elastic distortion, with atoms (1) to (4) remaining in contact; (d) if the atoms are hard spheres, the deformation cannot be the same for atoms (5) and (6) as it is for atoms (3) and (4).

well. This again cannot be calculated exactly, so that the problem is very difficult. However, instead of relying on calculations from first principles, we can take the elastic constants of various metals into account quite satisfactorily by introducing parameters given by experiment.

We can provide a *qualitative* explanation of some of the elastic properties observed in real solids by using the model of a pair of atoms. Figure 4.6, for example, can be used to predict the elastic behaviour of a pair of atoms, the modulus of elasticity being the same for compression and extension. If thermal vibrations are neglected, i.e. at 0 K, the modulus is given by the second derivative of the potential energy. This increases as the potential energy minimum becomes deeper, and thus as the cohesive energy of the solid becomes greater. This can be seen from Fig. 4.6 if we use the fact that the width of the potential well does not vary much from one element to another. It is therefore normal for solids with high melting points to have high Young's moduli. Thus, using a very simple argument, we are able to explain the correlations appearing in the experimental data given in Table 4.1.

Another fact related to atomic structure is the anisotropy in the elastic properties of a crystal. Even when there are many symmetry elements, the relative arrangement of neighbouring atoms is not the same in all directions: it is thus natural that Young's modulus should vary with changes in the direction of the tensile stress with respect to the crystal axes.

In its normal form, a metal is a collection of small crystals or grains with different orientations all adhering to each other. If the crystal is anisotropic, the individual grains suffer different strains, yet under stress their mutual adhesion is faithfully maintained. A grain within the polycrystal does not therefore suffer the same strain as it would if it were isolated. It can be seen from this that it would be very difficult to analyse the elastic strain of a polycrystal from the properties of a single crystal. We might expect a polycrystalline metal to be isotropic on the average if the grains were very small and had completely random orientations. The elastic parameters of the metal would then be the 'averages' of those of a single crystal. However, if such calculations are to be feasible, they must rest on simplifying and somewhat arbitrary assumptions.

When the metal has been subjected to cold-working, as in a rolled sheet or a drawn wire, the crystallites are no longer oriented at random: they have preferred orientations along characteristic directions such as the axis of the wire or the plane of the sheet. Under these conditions, the mean values of the elastic parameters of the individual grains will depend on direction. Thus, in a thin sheet, the 'longitudinal' Young's modulus (i.e. in the direction of rolling) is different from the 'transverse' Young's modulus. Here again, however, a theoretician can only make approximate predictions.

Secondary effects of elastic strain

The reversible change in the crystal unit cell (and in the neighbouring atomic motifs in amorphous materials (SM, p. 160)) caused by the application of an external force has consequences other than changes in shape. For example, the displacement of atoms in transparent materials, although small, modifies the conditions for the propagation of light waves: a cubic crystal or a piece of glass, both isotropic in the free state, become birefringent under compression.

In a quartz crystal, whether under tension or compression, the centroids of the positive and negative charges in the unit cell, which coincide in the free state, are separated in the stressed crystal and it thus becomes electrically polarized. The electric moment, like the strain, is proportional to the applied mechanical stress. This is the origin of **piezoelectricity**, whose important applications were mentioned in a previous chapter (p. 62).

Anelasticity

Elastic strain is due to the relative displacement of atoms, but the motion cannot be strictly instantaneous. Furthermore, since the atoms are in contact, their displacement involves frictional forces, which absorb energy. Real phenomena are thus more complicated than the description we have given so far: the elastic strain does not depend only on the force applied, but also on time, and the motion of the solid is damped to a greater or lesser extent. These effects give rise to what is called **anelasticity** and we describe several ways in which it manifests itself.

1. If we apply a tensile stress σ slowly and gradually to a specimen of length l, the elastic strain is Δl; but if it is applied abruptly, the instantaneous strain is less than Δl and it reaches this value only after a certain time delay (Fig. 4.8). This is known as relaxation, i.e. an evolution towards a more stable state. The rate of change varies from one solid to another and is characterized by a relaxation time τ.[†] In the same way, when the stress is removed, the solid does not regain its initial length instantaneously but only after a decay time characterized by the same parameter τ.

2. When a solid is subjected to a sinusoidally varying stress, the induced strain also varies sinusoidally at the same frequency. The system is similar

[†]The time τ is the time needed to reach two thirds (more precisely $1 - 1/e$) of the total relaxation according to the exponential law:

$$\Delta l(t) = \Delta l(0) + [\Delta l_{\infty} - \Delta l(0)]\left[1 - \exp\left(-\frac{t}{\tau}\right)\right].$$

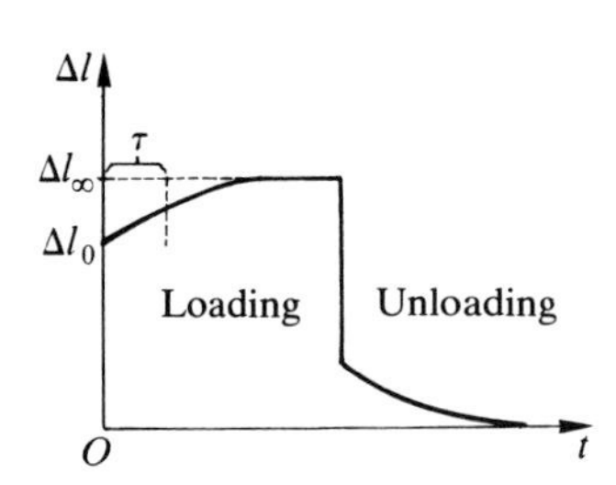

Fig. 4.8. Anelasticity: the variation of strain on loading and unloading.

to a damped harmonic oscillator. Because of the delay between the cause (the applied force) and the effect (the atomic displacements), the phase of the forced oscillation lags behind that of the external agent. It follows that the vibrating system absorbs energy, which is transformed into heat. This effect is known as **internal friction**.

In a torsional pendulum, the wire on which the weight is suspended is elastically strained by the torque. If the weight is set free, it oscillates with an amplitude that decays in time: the dominant factor in the damping is the internal friction in the elastic deformation of the twisted wire, provided the experiment is carried out with care so as to avoid mechanical friction.

3. When the frequency of the applied force varies, it is observed that the energy dissipated by the induced strain passes through a maximum when the frequency is $1/\tau$. This enables the relaxation time of a solid to be determined experimentally by studying the oscillations produced by a periodic stress of variable frequency.

Because the relaxation time depends on temperature, there is another method of determining τ. If the solid is subjected to a stress at a fixed frequency and variable temperature, the temperature at which the damping is a maximum can be measured: at this temperature, τ is the reciprocal of the forcing frequency.

We now attempt to explain these phenomena on an atomic scale, where the internal friction originates. The complexity of the problem arises from the fact that several different causes may be involved, so that there might be several different relaxation times for a given solid.

In order to show the variety of the atomic phenomena that cause anelasticity, we need describe only two very different sources of internal friction.

The effect of dislocations

We know that any real crystal contains dislocations. Texts on the atomic structure of solids describe the geometry of these defects (e.g. SM, p. 110) and we shall see later in this chapter how much they are involved in plastic deformation. Along dislocation lines and up to several atomic distances away, the atoms are considerably displaced from their equilibrium sites in the lattice. When the crystal is elastically stressed, the unit cells in the 'sound' parts of the crystal are only slightly deformed, but the atomic displacements along dislocation lines may be greater. As a result, the dislocation line can be moved within the crystal.

Suppose, as often happens, the dislocation line is 'pinned' at two fixed points one or two micrometres apart. The pinning is due to impurity atoms or intersections with other dislocation lines. Under the action of the force, the line is stretched between the fixed points just as it would be if it were an elastic cord fixed at the points. When the external force is removed, the dislocation line returns to its initial position under the action of internal elastic forces. When the crystal is subjected to an alternating force, the dislocation line vibrates between the pinning points: the resultant motion of the atoms is large enough to cause the vibration to absorb energy: this is one cause of the internal friction.

The effect of interstitial diffusion

Consider a crystal of α-iron (with a body-centred cubic unit cell) containing carbon. The iron atoms occupy all the lattice sites and the carbon atoms are added where there is most space: at the centre of the cube edges (Fig. 4.9). In spite of their small size, the carbon atoms distort the lattice and they are only 'tolerated' if the carbon concentration is very low, so that only a small proportion of the interstitial sites are occupied.

If the crystal is subjected to a tensile stress below the elastic limit parallel to the **c**-axis, the **c** edges will expand and ***a*** and ***b*** will contract. The carbon atoms have less space on the latter edges than on the ***c*** edges. Now the carbon atoms are not rigidly fixed to a site: they can jump from one position to an adjacent one and, if this is a more favourable site, will tend to remain there. The increasing occupation of the ***c*** edges and the removal of carbon atoms from the ***a*** and ***b*** edges produces a relaxation of the crystal, which extends until equilibrium is reached. Conversely, when the

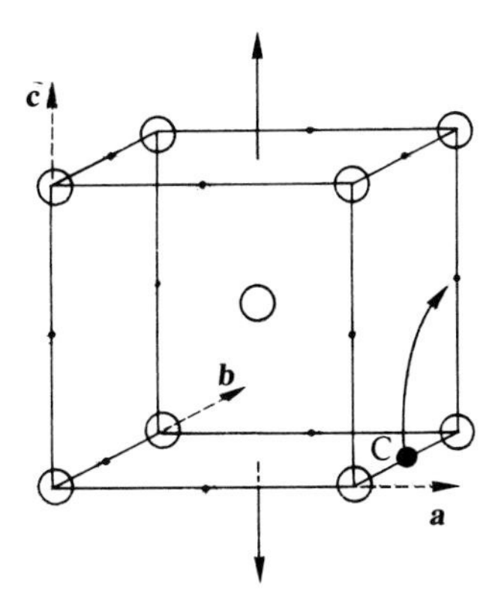

Fig. 4.9. The diffusion of carbon atoms in α-iron during an elastic tensile strain.

stress is abruptly removed, the carbon atoms gradually fall back into a random distribution which is the same for all three cube edges: the initial dimensions of the crystal are gradually restored.

Such 'directed' diffusion of carbon atoms in the iron delays the deformation under stress and is therefore a cause of internal friction. The relaxation time τ (or the frequency $1/\tau$) is a function of the mobility of the carbon atoms in the iron and, since this depends strongly on temperature, it can be seen that τ also varies with temperature in this case, as it does in general. The value of τ for iron–carbon alloys is of the order of 1 second at ordinary temperatures.

The mobility of atoms in the iron lattice varies from one element to another: thus, nitrogen is also a cause of internal friction but with a different relaxation time from that due to carbon. This gives us an understanding of how measurements of internal friction can identify the nature of the interstitial atoms present in the iron.

It is remarkable that simple macroscopic experiments like the damping of a torsional pendulum can yield an assessment of the probability that interstitial atoms will jump from one site to another in the lattice.

Plastic deformation

Ductile and brittle solids

When the tensile stress applied to a metal rod exceeds the elastic limit, it suffers a 'permanent set' or permanent extension after the stress is removed: this is the simplest manifestation of plastic deformation. In general, under the action of a system of sufficiently large forces, a solid can undergo a definite change of shape while remaining a coherent mass and very approximately preserving its volume. Thus, a cylindrical rod of cross-section S and length L passed through a drawplate can be transformed into a wire of smaller cross-section S/n and of length nL. A thick ingot passed through a rolling mill becomes a thin sheet with a large surface area.

Such behaviour, known as **plastic deformation** or **ductility**, is not typical of every solid. There are some that break suddenly without being deformed: these are said to be **brittle**. We shall be studying fracture properties in a later section.

For a given solid, ductility increases with rising temperature, sometimes appreciably. Man discovered long ago that iron can be shaped much more easily by forging at red heat (over 500 °C), whereas it is very difficult to work when cold. Similarly, plastics are very soft above 100 °C and quite hard at room temperature. A crystal of magnesia (MgO) is brittle at room temperature, but becomes ductile above 1200 °C.

To a first rather crude approximation, the two contrasting properties, ductility and brittleness, can be related to the nature of the interatomic bonds. Even if such a simple idea does have to be modified in a number of ways, it is important to make the point that crystals with metallic bonds are ductile, while other types, ionic†, covalent, and molecular, are brittle. It is well known that, at room temperature, a copper wire is easily bent while it is impossible to deform inorganic crystals or crystalline minerals: a piece of rock salt or a sugar crystal, for example.

Apart from metals, there are organic solids whose ductility is of great industrial importance: we call them, very appropriately, **plastics**. They are high polymers, i.e. having long-chain molecules assembled in the solid in a very imperfectly ordered state. The structure cannot be regular since the individual chains are flexible and have no well-defined shape. Their great ductility at moderate temperatures (about 100 °C) is due to the fact that the polymer chains become deformed and slide over each other. A description of the behaviour of plastics under the action of external forces can be found in texts dealing more thoroughly with their structure (e.g. SM, p. 174). We shall not be dealing with the subject in detail in spite of its practical importance because of the great complexity of the structures involved. We restrict ourselves to the ductility of metals, since only for very simple structures can we develop suitable atomic models.

Ductility from the macroscopic point of view

The main facts are as follows:

The solid changes shape without an appreciable change in volume.

In tensile tests, where the metal is subjected to a single stress along the axis of the specimen, the maximum extension varies between a few per cent and 50 per cent from one metal to another. It is thus considerably larger than the elastic strain, which is of the order of 0.1 per cent. The permanent set increases steadily with the applied stress up to the point at which the specimen begins to suffer damage (Fig. 4.10). Narrowing then occurs over a small section, an effect known as **necking**, which causes a local increase in stress; finally, the specimen breaks at the point where necking has occurred. The limits of the ductile state depend on the type of stresses imposed and they are therefore not an intrinsic property of the metal. For example, by repeated passages through a rolling mill, a sheet can be reduced to a tenth of its initial thickness, whereas in a tensile test the same metal cannot be stretched by more than 50% without failure.

†Except at high temperatures or with very pure single crystals.

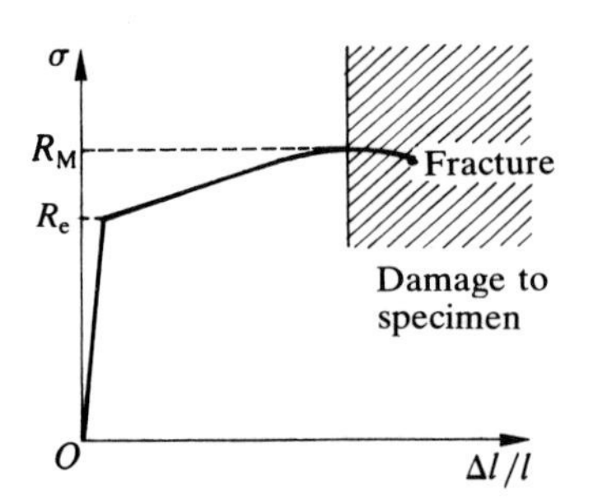

Fig. 4.10. A tensile curve: the relationship between stress and strain.

When the stress is removed after a certain extension, the metal has different mechanical properties from those it possessed before being strained. If it is subjected to a new tensile test (Fig. 4.11), its strain is elastic up to the maximum stress it has experienced: its elastic limit has thus been increased, sometimes considerably. For example, steel with an initial elastic limit of 1000 MPa may, after being drawn into a wire (a piano wire say), have an elastic limit of 4000 MPa. The plastic deformation has increased the strength of the metal, which can now withstand greater operational stresses without alteration. In short, the metal has been **hardened** and the process is known as **work-hardening**. The importance of such a simple method of improving the properties of metals can easily be appreciated: in fact, the majority of metals used in construction are in a work-hardened state.

The role played by metals in the civilized world stems largely from their ductility. A rough ingot can be given a shape approximating to the one required and we can then machine this material (described as semi-finished) into components which can withstand operating loads up to an appreciable fraction of those to which they were subjected during manufac-

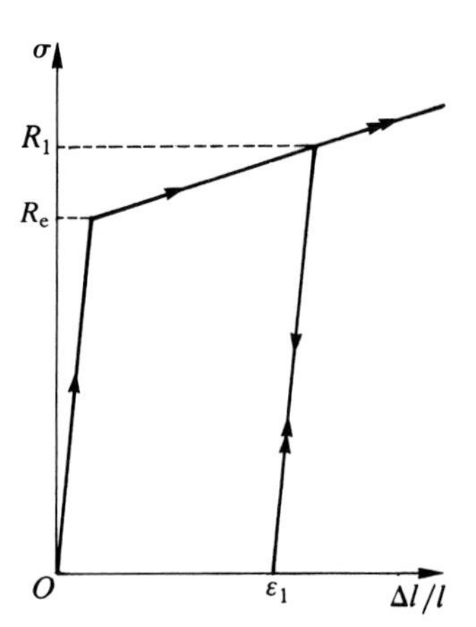

Fig. 4.11. Work-hardening. First experiment: for a stress R_1, the strain is ϵ_1. Second experiment: the elastic limit has become R_1, which is greater than R_e.

ture. In contrast to that, think of a non-ductile material such as stone, where the slab needed for a building, say, can only be cut directly from the rough block extracted from a quarry.

Slip in metal crystals

To throw some light on the atomic mechanism of plastic deformation, we start from some simple empirical facts of general validity.

In the first place, whatever the change in the external shape of the metal, even if it is very large, the material remains crystalline with the same crystal structure. To put this more precisely, the great majority of atoms after deformation are surrounded by the same arrangement of neighbouring atoms as before. For example, X-ray diffraction patterns show that copper, whatever deformation it experiences, is always crystalline with a face-centred cubic cell of side 0.36 nm.

The crystal grains, on the other hand, are affected a great deal: the deformed metal is still a polycrystalline aggregate, but it now consists of grains so fine that, for large strains, they are no longer visible under the optical microscope, being much smaller than a micrometre across. The large grains of the unstressed metal have been fragmented into a large number of crystallites with various orientations different from those in the initial grain. The new orientations are not randomly distributed: a given crystal axis in the various grains generally has a preferred direction that depends on the mode of deformation: this gives the solid a **texture** (SM, p. 126).

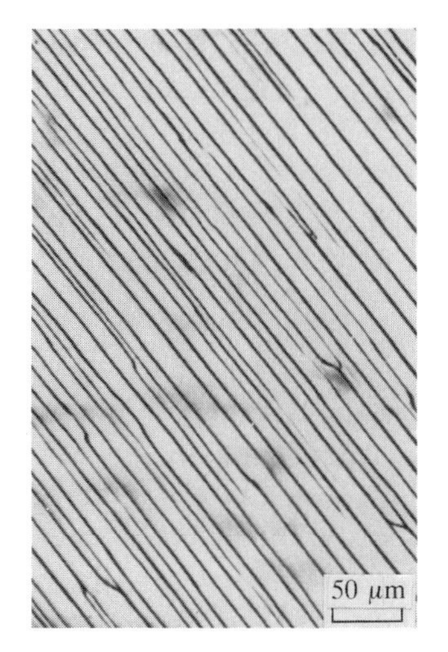

Fig. 4.12. Slip lines on an aluminium crystal (by kind permission of G. Wyon).

The most important feature, however, is the existence of **slip planes** in the plastically deformed metal. We start with the metal in an equilibrium state brought about by a long anneal. This has made the crystal as nearly perfect as it can be, and the crystals are large enough to be easily visible on a polished surface after it has been chemically etched. When the metal is subjected to a small plastic deformation, e.g. by a tensile stress, a series of fine parallel straight lines appears on the surface (**slip lines**: Fig. 4.12). The directions of these lines differ from one crystallite to another, showing that they are related to the specific orientations of the individual crystallites.

It can be shown that the slip lines are parallel to the intersection of a family of lattice planes with the surface. In a face-centred cubic metal, for example, they are planes normal to the cube diagonal (SM, p. 59). An examination of the surface at a greater magnification under an electron microscope reveals that a slip line marks out a step on the surface with a height that is variable but always very small, less than a micrometre (Fig. 4.13). A **slip band** can be resolved into a group of slip lines.

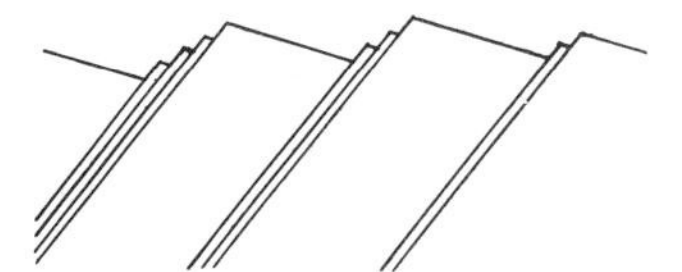

Fig. 4.13. Diagrammatic representation of slip bands resolved into groups of closely spaced slip lines.

These observations are explained by the **slip model**. A cubic crystal can be regarded as resulting from the close-packed stacking of planes of atoms, each plane consisting of a hexagonal array (SM p. 117): this is a structure typical of a metal crystal. For the stacking to be really close-packed, each layer must be so placed that its atoms lie in the hollows between the atoms of the previous layer. However, because the crystal has a regularly repeating period, one layer can be placed on another in an infinite number of equivalent ways. Two of these are shown in Fig. 4.14, and it can be seen that, although the external shape is different, the local atomic structure will remain unchanged. One layer will **slip** or **glide** over another when all the atoms in the first jump from one hollow to a neighbouring one. Such a process produces a strain with two of the characteristics of the plastic deformation of metals: a change in external shape and no change in the volume or the crystal structure.

In a stack of hexagonal layers, there are three equivalent slip directions at 120° to each other. The minimum step is equal to the distance between the atoms in contact and the slip in any one direction is equal to an integral number of steps. Moreover, a face-centred cubic crystal can be considered as a close-packed stacking of hexagonal layers in four different ways (these being normal to the four diagonals of the cube). For this crystal, therefore,

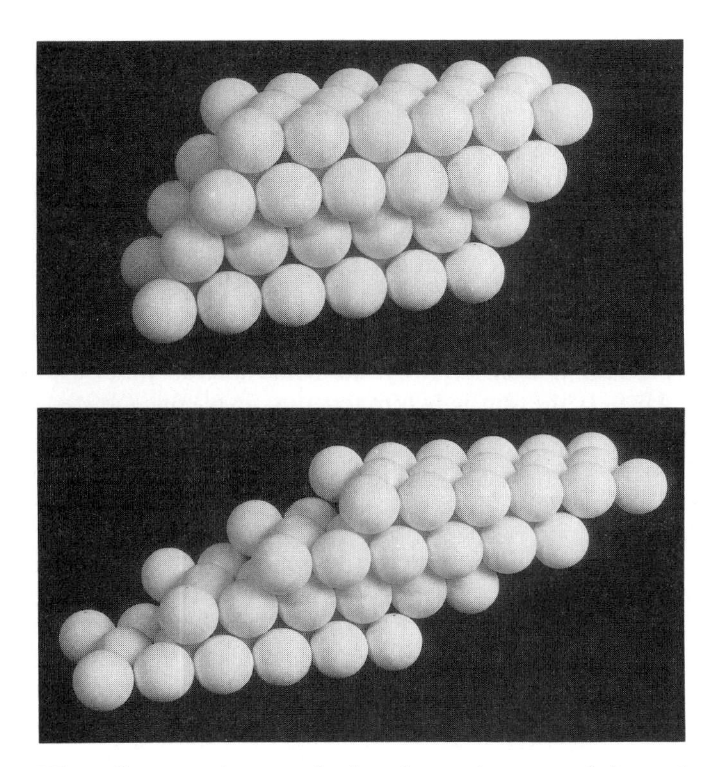

Fig. 4.14. When slip occurs between lattice planes, the external shape changes but the crystal structure remains the same.

there are twelve slip systems in all: each leaves the atomic structure intact, but it is easy to see that combinations of these various slip systems can lead to many different changes of shape for the crystal. In fact, it can be shown geometrically that a crystal can be given any shape whatsoever, all of the same volume, by the action of only five of the twelve possible slip systems.

To initiate slip in a crystal in stable equilibrium, a shearing force parallel to the slip direction must be applied. The force required is that necessary to make the atoms of a lattice plane leave the hollows in which they are lodged and take them through a non-equilibrium state until they fall once more into another set of hollows, when the crystal is once again in stable equilibrium but with a new configuration.

For such a process to be possible, it must not need too great a supply of energy to the crystal. This condition is satisfied by metals because the interatomic bonds are undirected: their only effect is to make the whole atomic assembly as close-packed as possible. In the slip process, it is true that the configurations intermediate between the two stable ones of the same volume do involve a departure from the closest possible packing, but it is not very large and therefore does not require too much extra energy.

On the other hand, slip is difficult in covalent crystals because the bonds change direction during the process. The bonds have to deviate from the arrangement determined by the covalency and that cannot occur without the supply of considerably more energy than in the case of metals. For example, silicon shows no plastic deformation at room temperature and only begins to show it above 400 °C, when the thermal vibrations of the atoms have sufficient amplitude.

Similarly, slip in ionic crystals could upset the local electrostatic equilibria and would thus require a larger input of energy.

Two ways in which slip can occur can be envisaged. In homogeneous slip, every plane is translated by the same vector with respect to the adjacent plane. If the slip is heterogeneous, it is concentrated in certain planes or groups of neighbouring planes called bands, while the crystal between two slip bands remains intact. The macroscopic effect on the external shape is almost the same as in the first case if the bands are close enough.

Experiment shows clearly that the process is *heterogeneous*, since distinct slip lines can be seen (Figs 4.12 and 4.13) separated by patches of smooth surface with widths of the order of at least a micrometre: these contain several thousand lattice planes which have suffered no slip.

Thus, the slip model, based soundly on experimental observations, provides a *qualitative* explanation of plastic deformation and of the ductility of metals. But is it possible, using this model, to arrive at a *quantitative* explanation of ductility?

An attempt at a theoretical evaluation of the elastic limit

Our aim is to predict the minimum stress that needs to be applied to a metal to initiate plastic deformation. To simplify the problem, we shall consider an isolated single crystal of the metal and base our argument on a two-dimensional case, which may be unrealistic but is enough to reveal a fundamental difficulty.

Consider two rows of atoms in contact (Fig. 4.15) and in stable equilibrium, which means that they are stacked so that each atom sits in the hollow formed between two atoms of the other row. If a small shearing force is applied to the system, it starts to move the atoms from their equilibrium position but does not make them leave their 'hollow'. As soon as the stress is removed, the atoms return to their equilibrium positions, so that we are in the elastic region of strain.

By increasing the shearing force, the atoms eventually leave their hollows and are taken to a position directly over the atoms in the adjacent row: this is a position of unstable equilibrium. A very small increment of force in the direction of the initial stress makes the atoms fall into the set of

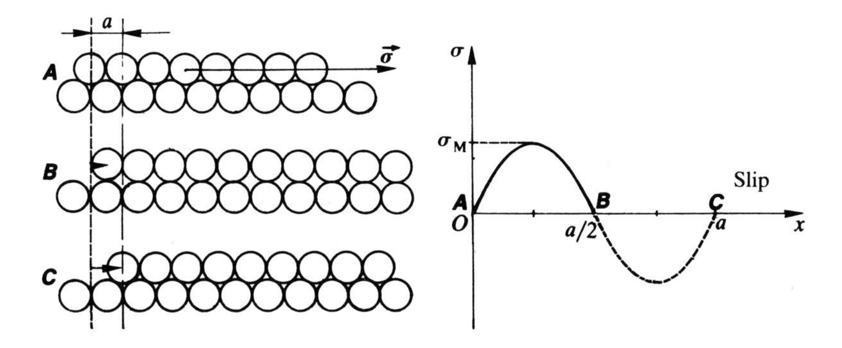

Fig. 4.15. Calculation of the theoretical elastic limit in a perfect crystal.

interstices adjacent to those they occupied initially. Slip has occurred over a distance equal to the atomic diameter. The elastic limit corresponds to the minimum force that had to be applied in order to produce a slip of one step.

A simplified model of this sort enables us, if not to calculate accurately, at least to make an estimate of the elastic limit in terms of the modulus of elasticity (Box 19).

Box 19

Theoretical calculation of the elastic limit

The elastic limit is reached when two planes of atoms can slip with respect to each other, thus producing an irreversible strain in the material.

Suppose therefore that there is a stress σ tending to displace all the atoms in a plane with respect to those in the plane immediately below it (Fig. 4.15). As long as this stress remains small, the displacements are also small compared with the interatomic distance and the stress needed to produce the strain is proportional to the displacement. However, this relationship soon fails when the displacement amounts to a significant fraction of the interatomic distance a. A special situation occurs when the displacement is $a/2$ and the unstable position shown in Fig. 4.15B is reached. The stress is then zero and, by however small an amount the position shown in Fig. 4.15B is exceeded, the plane of atoms is attracted naturally into a new equilibrium position (Fig. 4.15C), displaced by a from its starting position (Fig. 4.15A). A reasonable expression for the stress as a function of the displacement is given by the upper half of a sinusoidal curve:

$$\sigma = \sigma_M \sin \frac{2\pi x}{a}.$$

The maximum σ_M corresponds to the minimum stress that must be applied for slip of the planes to occur: this is the theoretical elastic limit. It can easily be related to the rigidity modulus: for small displacements, σ can be expressed approximately by:

$$\sigma = \sigma_M \frac{2\pi x}{a}.$$

Now, according to the definition of rigidity modulus (Box 18), the displacement of the upper plane with respect to the lower, when small, must be $x = \alpha d$, where d is the interplanar distance and α is the angle of shear of the lateral faces given by $\alpha = \sigma/\mu$ (p. 180).
We therefore have $x = \sigma d/\mu$, which gives

$$\sigma_M = \frac{\mu}{2\pi} \frac{a}{d}.$$

The interplanar spacing d is approximately equal to the atomic diameter a, so that we have:

$$\sigma_M \approx \mu/6.$$

In fact, it can be seen from Table 4.1 that there is a good correlation between the elastic limit and Young's modulus and therefore between the elastic limit and the rigidity modulus μ (p. 182), but the ratio σ/μ is much smaller than the value predicted by this theory.

The important feature of the calculation is the order of magnitude of the result. It predicts that the stress which must be applied to make the first slip lines appear is of the order of a tenth of the rigidity modulus.

Experimental values of the elastic limit

Consider the case of a tensile stress applied to a test specimen consisting of a single crystal (Fig. 4.16). The slip lines determine the orientation of the slip plane. The effective shearing force is equal to the projection on this plane of the axial tensile force and is therefore $F\cos\alpha$. The shear stress is the ratio of this projection to the area of cross-section of the specimen in the slip plane, $S/\sin\alpha$, and is thus $(F/S)\sin\alpha\cos\alpha$.

When the experiment is repeated with various crystal orientations, slip appears for different values of the tensile force, but the value of the stress

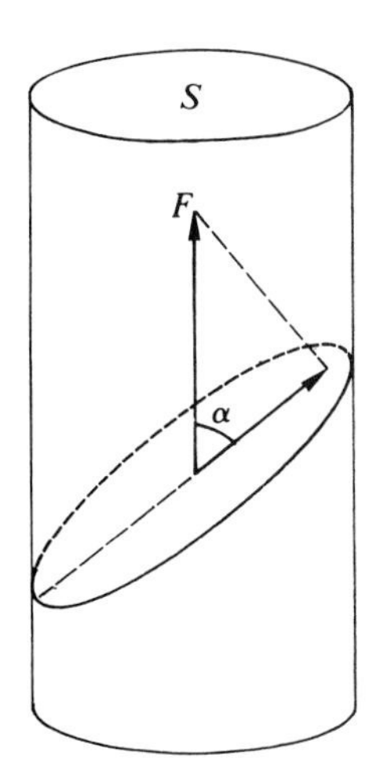

Fig. 4.16. Schmid's law: the stress in the slip plane is $F\cos\alpha/(S/\sin\alpha)$.

along the slip plane is the same (Schmid's law): it is this stress which must be compared with the calculated value for the onset of slip.

The highly significant result is that *the experimental value is between 100 and 1000 times smaller than the theoretical value*. Such a large discrepancy cannot be corrected by the mere refinement of a calculation that is so obviously imperfect. It is the model itself that has to be abandoned, a step that lies at the root of one of the main ideas in the physics of solids, at least as far as mechanical properties are concerned.

Our model is inadequate because it is based on the assumption of a perfect crystal, which simply doesn't exist.† All real crystals contain defects which play such an important role in mechanical properties that calculations starting from the perfect crystal model are no longer valid at all.

Dislocations: their role in plastic deformation

The defects mainly involved in plastic deformation are **dislocations**. These are described in detail in texts dealing with the atomic structure of solids (e.g. SM, p. 110) and methods for making them visible are also usually described. We review here the structure of a single type of dislocation, the edge dislocation, which is the easiest to represent diagrammatically.

A half plane is removed from a set of parallel atomic planes (Fig. 4.17) and the empty space left behind is filled by the relaxation of the crystal, its two halves coming closer together. The re-formed crystal is very nearly

†It is said (e.g. SM, p. 106) that silicon crystals as large as several cubic millimetres can be obtained without any defects, but silicon does not undergo plastic deformation at ordinary temperatures.

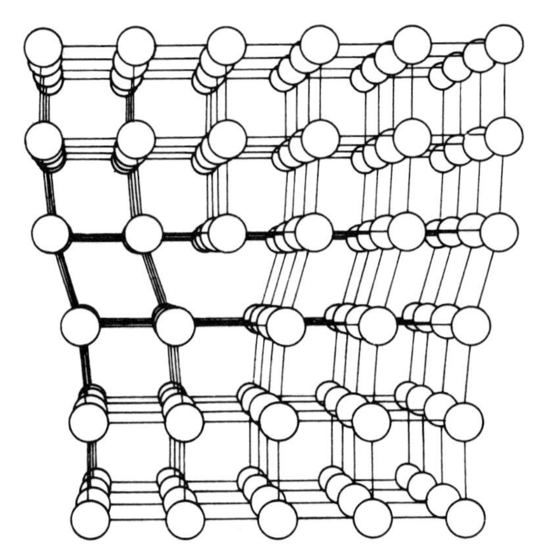

Fig. 4.17. The structure of an edge dislocation in a simple cubic lattice.

perfect, except inside a 'tube' surrounding the **dislocation line** which runs along the edge of the half plane of atoms and forms the core of the dislocation. The disturbance of the lattice becomes negligible compared with thermal agitation at about 5 nm away from the core.

In the model with two rows of atoms used previously, we can introduce something similar to a dislocation by removing an atom from one of the rows. Figure 4.18 shows that the arrangement of atoms in the two rows is normal except in a small group of atoms that are in an unstable state. If a shearing force is applied, the disordered group is displaced along the row until it reaches the end, where it disappears, leaving a step. If a second

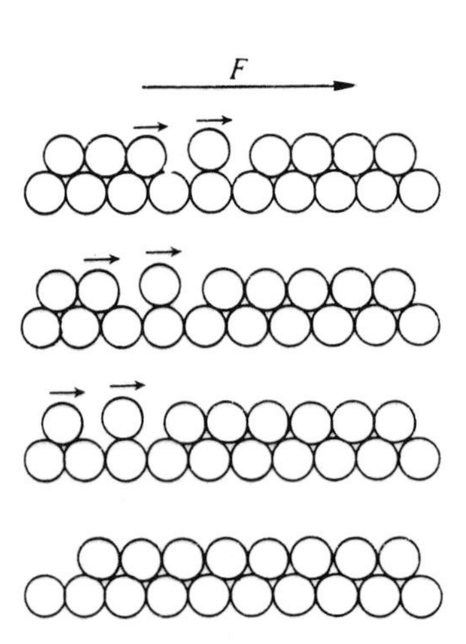

Fig. 4.18. The glide of a dislocation represented by a missing atom. Under the action of an external stress, the atoms around the dislocation slip to the right. The dislocation moves to the left and eventually disappears, leaving a surface step with a depth equal to one interatomic spacing.

dislocation is created at the beginning of the row, it will also slip to the other end and double the size of the step. A single dislocation crossing the system produces slip of one atomic step.

It is the same in a crystal: the displacement of the dislocation line along a lattice plane produces slip equal to a single interatomic spacing.

The essential feature of the slip process in a crystal containing a dislocation is that the shearing force needed to produce the slip is considerably reduced by the presence of the defect. Only a few atoms near the dislocation are displaced; and, what is more, half of them tend spontaneously to fall into their new position. It follows that very few atoms need be subjected to the force needed to change their position. The final deformation is the result of *successive* displacements of the atoms, whereas, in the perfect crystal, all the atoms *simultaneously* resist the shearing force when slip begins.

This explains why real crystals containing dislocation lines have elastic limits that are several orders of magnitude smaller than those calculated for perfect crystals.

It was in fact precisely this situation that led theoreticians to introduce the concept of a dislocation well before there was direct experimental proof of their existence. As we shall see, the model does not enable us to calculate the elastic limit from first principles but it does explain several experimental facts.

1. The **Bragg bubble raft**, described in many texts on defects in solids (e.g. SM, p. 110), is a two-dimensional model of the close-packed stacking of atoms. The apparently regular array of bubbles in this experiment can be observed to contain dislocations exactly like the edge dislocation shown in Fig. 4.17. If the raft is held between two parallel glass rods, they can be moved in such a way as to apply a shearing force to the array and during the resultant deformation dislocations can be seen slipping rapidly along the dense rows of bubbles.

2. In an electron microscope picture of a very thin section of a metal, the dislocation lines appear black. If the section is slightly deformed *in situ*, the dislocation lines can be seen to move and their motion can be filmed.

3. The variation of elastic limit with temperature is particularly large in body-centred cubic metals (SM, p. 92), which is of great importance technically: this property is related to the atomic structure in the dislocations core.

4. If a metallic crystal contained no dislocations, its elastic limit ought to be considerably greater. However, it is impossible in practice to prepare large metallic crystals that are perfect. In certain reactions, on the other hand, very fine metallic spikes are formed, several micrometres in diameter and several millimetres long, known as **whiskers**. These are single

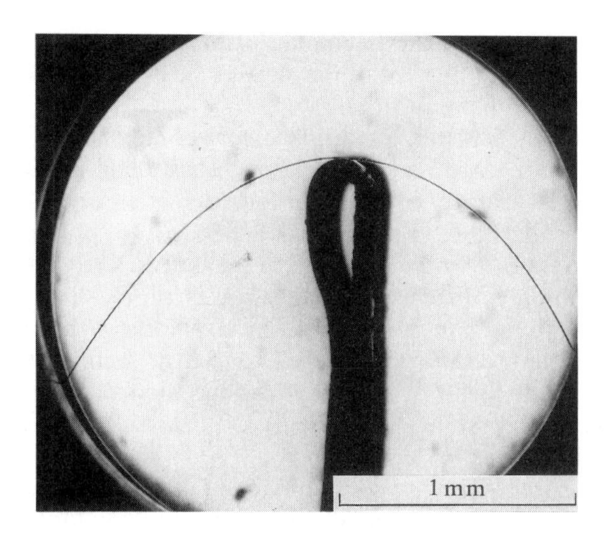

Fig. 4.19. A whisker (diameter 5 μm), fixed at both ends, is curved elastically by applying a force at its centre with a hook

crystals and the electron microscope shows that they contain no dislocations. It has also proved possible to bend whiskers under an optical microscope (Fig. 4.19) and to evaluate the maximum stress below which they remain elastic. The observed elastic limit is found to be much greater than that of the bulk metal and to approach the value for a perfect crystal.

Towards a quantitative theory of the ductility of real crystals

We can be confident that the key to the phenomenon of plastic deformation is the motion of dislocations in a crystal under the action of the applied force. However, the physicist wishes to be able to make quantitative predictions of the mechanical properties of metals, particularly as functions of temperature and, to achieve this, the model of the crystal with dislocations must be subjected to calculation. Many problems arise when this is attempted, some of which have not yet been completely solved. In this section, we describe some of the difficulties and show how far we can go in spite of them.

The origin of dislocations

According to our model, slip of one atomic step corresponds to the disappearance of one of the dislocations in the crystal. It might be concluded from this that the stage during which slip occurs easily must stop when there are no longer any dislocations in the crystal, producing an overall deformation that is in fact quite small. After that, the crystal could not be deformed any further without the application of a much greater external force.

This is not what is observed. Plastic deformation goes on until strains of several per cent have occurred, after which the increase in stress takes place more slowly but still continues (Fig. 4.10). Dislocations must therefore always be present, even though some disappear; it follows that **dislocation sources** must exist within the crystal to replace those that disappear and even to multiply the number that are already there. Observations under the electron microscope show that the number of dislocations increases rapidly with the amount of plastic deformation.

We recall that slip in a crystal is not homogeneous: over narrow bands, the slip may amount to several thousand interatomic distances, while intermediate blocks are not significantly deformed. This arises from the fact that the dislocation sources needed for large amplitude slip are not uniformly distributed throughout the crystal, and slip only occurs where such sources are located.

A mechanism by which dislocations could multiply was suggested by Frank and Read (1950). We shall not describe it here, but it is worth pointing out that the 'Frank–Read source' was eventually observed exactly as theoreticians had predicted several years previously (as has happened several times in the history of solid state physics).

Dislocation dynamics

Ductility relies on the displacement of dislocations in a crystal. The corresponding laws governing their motion in a force field must be known if we want to calculate, for comparison with experiment, the minimum stress for slip to occur, and to explain its increase when the metal is deformed (work-hardening) or when the temperature is lowered.

The elementary process is the **glide** of a dislocation in an otherwise perfect crystal along a lattice plane having a high density of atoms. The gliding speed cannot exceed that of sound (about $5 \times 10^3 \mathrm{m\,s^{-1}}$) since no interaction between atoms can be propagated any faster than this, and in fact the speed at which dislocations move is thought to be considerably smaller.

The force required to maintain this motion can be evaluated and is found to be much less than the experimental elastic limit. The situation is therefore as follows: for a perfect crystal without defects, the calculated value is too high, but a model involving isolated dislocations in an otherwise perfect crystal yields a value that is too low. It is clearly necessary to find some reason for the resistance to the motion of dislocations in a real crystal.

The cause might be the fact that a dislocation moves in a crystal that is not 'otherwise perfect' but is full of various other imperfections: grain boundaries in polycrystalline metals, other dislocation lines with their cores within which the lattice is distorted, point defects such as vacancies and, unless the metal is very pure, impurity atoms, either dispersed, or assembled in small grains of a second phase or in GP zones (p. 210: see also SM, p. 119).

The problem thus appears extremely complex. Firstly, it is clearly difficult to define a real structure of the crystal on which to operate, since the imperfections are, by their very nature, irregular. Secondly, the interaction of a dislocation with each of the other types of imperfection poses a problem for which there is no exact solution. How does the gliding dislocation line cut the 'forest' of lines normal to the slip plane? How does it move in the crystal structure of a solid solution, where the atoms replacing those of the matrix produce slight but very frequent perturbations? What effect does a variation of temperature have on the different interactions? And so on.

The difficulty with each of the problems is obvious: not only are there so many interacting atoms that it is impossible to know the exact positions of all of them but, as we have often said, the laws governing the atomic interactions in terms of their distance apart and their surroundings are not known with sufficient accuracy.

All these problems have received attention in recent decades and some progress has been made. It has proved possible to estimate the orders of magnitude of interactions for several types of defect and to show how the effects are combined in a real crystal. Nevertheless, a calculation of the macroscopic properties of a metal from first principles appears to be beyond us: even the tensile curve for a single crystal of a pure metal, which would be an extremely simple case, cannot yet be predicted.

We also expect the variations from one specimen to another to be large and unpredictable, since the defects in them might be very different and there is no non-destructive method of determining them in a given specimen. Experiment is unfortunately of little help to the theory, because it is seldom that the effect of a particular defect can be isolated. Measurements are made on specimens which are bound to contain many types of defect and it is almost impossible to extract data on a single type that could be used in a theoretical calculation.

Some qualitative applications of the theory of ductility

As a counter to the above somewhat pessimistic comments, we should instance some of the technologically important results obtained by metal physicists, whose ideas have enabled metallurgists to rationalize their observations and measurements. When their empirical research receives guidance in this way, it is possible for them to pick out the directions which have some chance of success and reject those that would only be a waste of time and money.

We illustrate these points with several important examples, starting from the very simple and rather vague idea that dislocations are easily moved in a perfect crystal and that they are retarded by anything that disturbs the order in the crystal. We shall show how it is possible to exploit the idea that mechanical strength depends on the number and nature of the structural defects.

Work-hardening

A metal crystal is never free of dislocations, however much care is taken in preparing it. The density of dislocation lines is specified by the length of the lines contained in unit volume and is measured in $\mathrm{cm\,cm^{-3}}$, i.e. in $\mathrm{cm^{-2}}$ (this is also the number of lines crossing unit area). After a long anneal at a high temperature, which enables the atoms to take up energetically favourable positions, the dislocation line density is of the order of $10^6\,\mathrm{cm^{-2}}$, which means an average distance between lines of about $10\,\mu\mathrm{m}$.

When plastic deformation begins, the original dislocations certainly move, but the predominant effect is the creation of a much greater number of new dislocations generated by localized active sources. The movement of these dislocations enables the deformation to continue and to increase. Some dislocations disappear in the process, but many more persist in the crystal, which thus becomes more and more imperfect. This is revealed by electron micrographs taken at various stages of deformation. Finally, in a metal that has been heavily rolled or drawn, the dislocation density can be as high as between 10^{10} and $10^{12}\,\mathrm{cm^{-2}}$, or an average of one line every

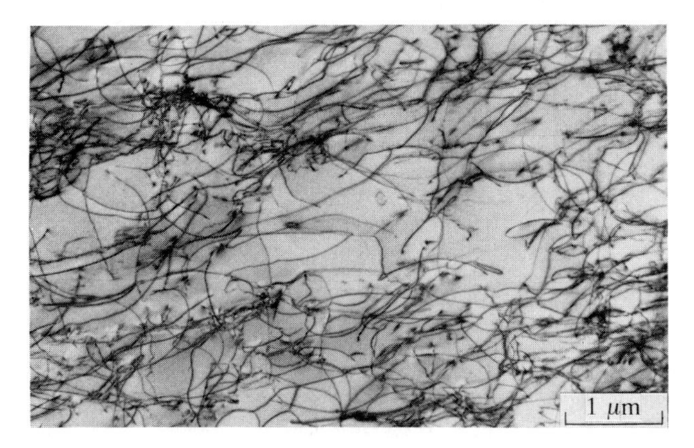

Fig. 4.20. Dislocations in work-hardened stainless steel (by kind permission of C. Donadille).

10 nm, although the dislocations are in fact non-uniformly distributed and form a highly tangled network (Fig. 4.20).

In the deformed metal, the dislocations have to move through a crystal with a dislocation density that increases as the strain becomes greater, and the motion of some will be blocked if the external stress is insufficient. There is certainly a greater resistance to the movement of dislocations and it also becomes more difficult for new sources to be produced.

This is the origin of **work-hardening**, i.e. the increase in the strength or hardness of a metal resulting from deformation. A metal crystal containing no dislocations would, of course, be very strong, but it is impossible to remove them all, so that this is not a feasible method of hardening. The other possible method is to impede the motion of the dislocations that do exist, and this is achieved in work-hardening through the interaction between dislocations whose density is very high.

It has long been known that work-hardened copper tubes (i.e. after cold working of the metal) are rigid while annealed copper tubes can easily be deformed. The difference between the two states can be strikingly demonstrated by quite simple experiments (e.g. SM, pp. 133–4). In the annealing operation, the small and very imperfect crystals of the work-hardened copper disappear and are replaced by larger and more perfect crystals: many dislocations disappear and others are grouped together in a way that is energetically more favourable. The decrease in the strength of a metal in the annealed or 'recrystallized' state is caused by a reduction in the dislocation density.

Effect of grain size

This is a general effect which is easily observed. For a given metal, an isolated single crystal is not nearly as strong as a polycrystalline aggregate of small contiguous crystal grains. The strength is also increased by a reduction in the average grain size. To give some idea of orders of magnitude: the elastic limit of copper doubles when the average grain size falls from 100 μm to 25 μm.

These effects are quite easy to explain. In a single crystal, the dislocations move through the whole specimen during slip and eventually disappear at the free surface, forming steps. In the polycrystal, on the other hand, a slip plane in one grain does not correspond to a slip plane in an adjacent grain, so that the moving dislocation cannot pass the grain boundary. The two grains remain contiguous and so the dislocation cannot disappear and leave a step on a free surface. The dislocation is therefore blocked and, in turn, blocks the following dislocations. This causes them to pile up against the grain boundaries, an effect that can be observed in electron microscope pictures (Fig. 4.21). A calculation of the interactions between dislocations has even made it possible to explain the variation in

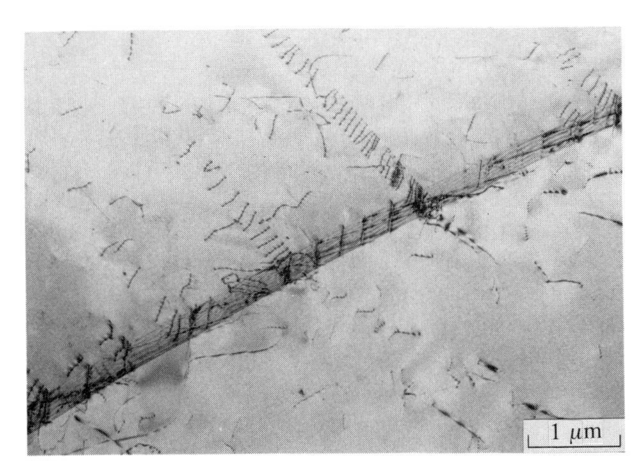

Fig. 4.21. Dislocation pile-up at a grain boundary (electron micrograph taken with a 1 million volt instrument) (by kind permission of C. Donadille).

the interval between successive blocked dislocations with increasing distance from the boundary where pile-up occurs.

Dislocations can also be created at the grain boundaries and there will thus be more of them as the total surface area of the grain boundaries increases, i.e. as the grain size is reduced. Many of these dislocations remain blocked in the grains, so that the dislocation density is greater with smaller grains and the strength of the metal is increased.

A large number of measurements have shown that this effect exists. For a metal of given composition and a given degree of strain, the yield stress (at which plastic deformation begins) increases as the grain size decreases according to the Petch equation (Fig. 4.22, which also shows that the effect of work-hardening adds to that of grain size).

This example typifies the state of our knowledge in metal physics: the effect can be explained qualitatively, but the law expressing the effect is empirical.

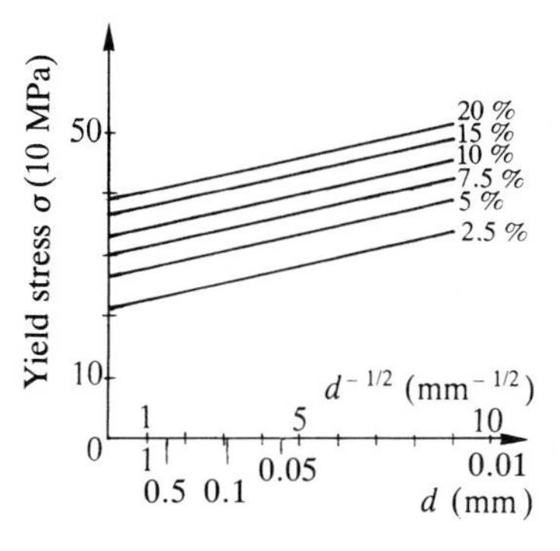

Fig. 4.22. Relationship between the yield stress of mild steel and grain size for various strains, illustrating Petch's equation: ($\sigma = A + Bd^{-1/2}$).

A technically important consequence of all this is that the quality of a metal can be improved by giving proper consideration to ways of controlling grain size. This is how high performance steels are produced, and it demands detailed and thorough research to determine the optimum conditions for heat treatment (temperature and time), as well as very strict monitoring and control of the industrial manufacturing process. It has also been found that the addition of certain elements to steel, such as niobium in very small proportions (0.04 per cent), produces a smaller grain size. This is because the added elements form compounds that crystallize as very fine grains distributed throughout the bulk of the metal. These then fix the grain boundaries, so that the grain size cannot increase during annealing.

Ductility in alloys

Alloys contain several constituent metals, which are generally mixed in the liquid phase to obtain the final product. Their mechanical properties, like those of pure metals, are affected by work-hardening or grain size. However, there are in addition other effects related directly to their composition.

Solid solutions

The simplest type of alloy consist of a single phase: it is then a **solid solution**. There are two main categories of solid solution: **substitutional solutions** in which the dissolved atoms replace those of the basic metal at equivalent lattice sites in a single crystal; and **interstitial solid solutions**, in which the added elements are lodged in interstitial sites between the normal lattice sites of the solvent element. The latter type can only be formed if the added elements have atomic diameters that are small compared with that of the solvent element and if the proportions of the added elements are not too high. The most important example of this second class is that of an interstitial solution of carbon in iron, as found in steels.

The dissolved atoms deform the crystal lattice of a solid solution on an atomic scale by amounts that depend on their nature and concentration and on the type of solid solution. In substitutional solutions, the lattice distortion arises because of the difference in the atomic diameters and the interactions of substituted atoms with neighbours. In interstitial solutions, the distortion is greater since the interstices are always markedly smaller than the atoms they accommodate. During plastic deformation, the dislocations are therefore moving through a medium with a large number of irregularities, which impede the motion to a greater extent than in a pure metal.

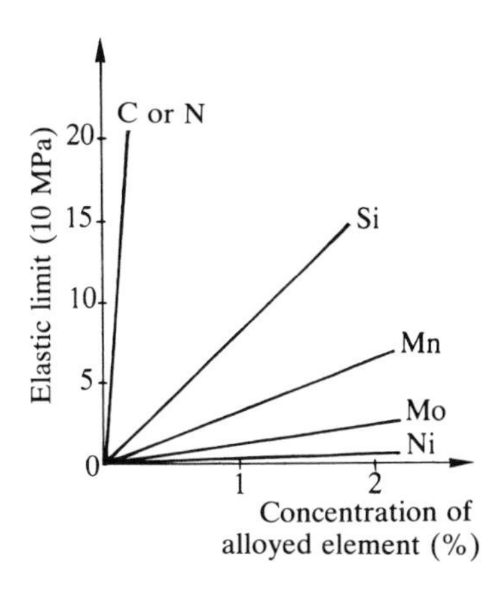

Fig. 4.23. The effect of various elements dissolved in iron on its elastic limit: the variation of the limit with the concentration of the added element.

We thus predict that the elastic limits of homogeneous alloys will be higher than those of pure metals and this turns out to be the case: a gold alloy containing 10 per cent copper is three times harder than pure gold. The increase in the elastic limit becomes greater as the lattice distortions increase. Figure 4.23 shows that, for iron alloys, this increase is proportional to the concentration of the added element and that, for the same atomic concentration, it varies from one element to another. Finally, it is considerably greater for interstitial solutions than for substitutional solutions.

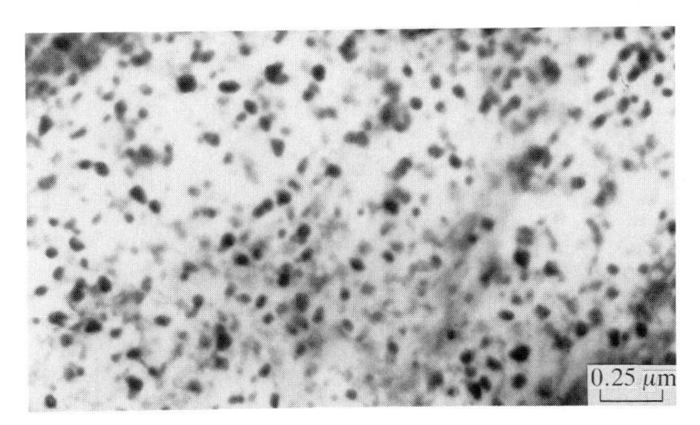

Fig. 4.24. Microprecipitates of Fe_2Mo in an Fe–Mn–Co–Mo alloy (by kind permission of C. Servant).

Alloys containing precipitates (precipitation hardening)

We consider two-phase alloys in which the principle phase (or matrix) is predominant and the precipitated phase is dispersed in the form of numerous very fine grains incorporating a high proportion of the atoms added to the principal constituent (Fig. 4.24). The deformation of such an

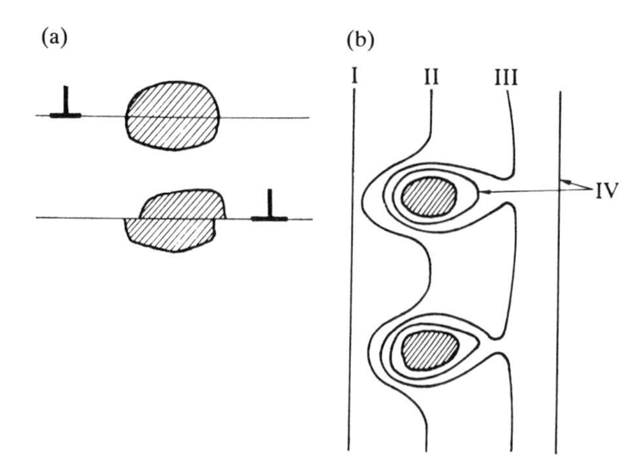

Fig. 4.25. Encounter between a dislocation line and precipitate grains: (a) shearing of a small grain; (b) the line bypassing grains (four successive positions of the line are shown).

alloy takes place via slip in the matrix. When mobile dislocations encounter the precipitated phase, these small, hard, not very ductile grains become serious obstacles: the force produced along the dislocation line may be enough to shear the small grain, or the grain may be bypassed by the dislocation. Figure 4.25 shows both possible processes. In either case, the presence of small grains included in the matrix strengthens the metal considerably. This is one of the best methods available to metallurgists for hardening metals.

Alloys with GP zones: the duralumins

Some alloys are in a state intermediate between those just described. The dissolved atoms are not distributed uniformly as in a homogeneous solid solution and they are no longer separated from the matrix in the grains of a second phase. Instead, the crystal contains Guinier–Preston or **GP zones** (SM, p. 119). These are very small, of the order of 10 nm across. The concentration of dissolved atoms in these zones is very high compared with the average concentrations; the structure nevertheless remains close to that of the matrix, but the lattice is highly distorted. The zones may have a particular shape: very thin platelets, parallel to the faces of the cubic unit cell if the matrix in which they are formed has a cubic lattice.

A typical example is an alloy of aluminium and copper (5 per cent copper), which at high temperatures (550 °C) is a homogeneous solid solution. If the alloy is suddenly cooled by quenching, it remains a solid solution, but it is no longer in a stable state. This is because, although the

5 per cent of copper is soluble in the matrix at the high temperature, only 0.5 per cent is soluble at room temperature: the solid solution preserved by quenching is **supersaturated** and the copper atoms therefore tend to separate out from the aluminium. The equilibrium structure would consist of two phases: an aluminium matrix, with precipitates rich in copper (in this case, these are crystals with a composition Al_2Cu).

If we now maintain the quenched alloy at room temperature, this equilibrium state cannot be reached because the copper atoms have to diffuse through the aluminium matrix over distances of the order of several micrometres. This is impossible at room temperature because thermal agitation is too small, and so the copper atoms become grouped together in the aluminium crystal in a large number of very small zones: the atomic migration needed is only over a few nanometres and takes place slowly (in one day). The zones are revealed by X-ray diffraction and electron microscopy. The solid solution has become heterogeneous on a scale of 10 nm and, in these zones, is in a metastable state different from the equilibrium state but more stable than the homogeneous solution. If the alloy is annealed at a temperature of about 300 °C, the mobility of the atoms is sufficient for equilibrium to be established: the precipitation of the copper in the Al_2Cu phase is complete in a few minutes.

The changes in structure are accompanied by a change in mechanical properties. After the quench, the alloy is not very strong but it becomes harder when the GP zones are formed by ageing at room temperature. It becomes soft again after an anneal, when the precipitates appear.

Such a complicated set of changes is well explained by theory. In all three states, there are local deformations of the crystal which impede the movement of dislocations. In the homogeneous solid solution, distortion around the isolated copper atoms is small: the alloy is therefore not very strong, although it is much stronger than pure aluminium. In the precipitated state, each Al_2Cu grain is a considerable obstacle: but the precipitates are a long way from each other and, between them, the crystal is sound and dislocations can move easily. In this state, the metal is not very strong and its properties revert approximately to those of the homogeneous solution. When the crystal contains GP zones, on the other hand, the lattice distortion is large, firstly because the zones are very numerous and therefore very close, and secondly because the crystal is highly distorted inside the zones. The overall result is that the propagation of dislocations is made particularly difficult and the alloy has hardened.

The Al–Cu alloy is typical of the **structural hardening** of certain alloys which was discovered empirically (Wilm, 1911). Thanks to the dislocation model, we now understand the behaviour of these remarkable alloys and it has been possible to make continual improvements in them. During the few hours after a quench, the highly ductile alloy can be worked easily.

Then, through spontaneous ageing and without any change in external shape, the alloy hardens so that it can withstand large operational stresses without deformation. Not only that, but when the basic constituent is aluminium, the alloy is light: for equal weights, duralumin structures have a performance close to that of steels. The alloy used in Concorde, for example, containing 2 per cent copper, 1.5 per cent magnesium, and 1 per cent nickel, has an elastic limit of 450 MPa up to an operating temperature of 120 °C.

Steels: martensitic transformations

Steels are alloys of iron and carbon, often with the addition of other metals in various proportions. For all of us, steel is the typical example of a hard and very strong metal. Hundreds of varieties with different compositions are at present commercially available, possessing properties that cover an extremely wide range because they depend not only on the composition but also on the treatment to which they have been subjected. Hardened steel, for example, has been used for thousands of years in the production of efficient weapons and reliable tools.

How can theory help us in understanding all the various properties of steels? We make no attempt here to provide a general answer to this question because of the wide variety that exists in their composition and treatment: our description is instead confined to an outline, with the help of structural models, of the effects that a phenomenon characteristic of steels has on their ductility: namely, the **martensitic transformation**.

Pure iron occurs in two forms with different crystal structures: elements behaving in this way are described as allotropic and their various forms as allotropes. Above 911 °C, iron crystallizes with a face-centred cubic unit cell (γ-iron), while below that temperature it is body-centred cubic (α-iron) (Fig. 4.26). This is still the situation in steel, although the transition temperature then varies with the composition of the alloy.

Consider a very simple steel, an Fe–C alloy with 0.5 per cent carbon. When in equilibrium at 950 °C, it crystallizes in the γ form: the carbon is distributed uniformly in an interstitial solid solution known metallurgically as **austenite**. This is now brought suddenly to room temperature by quenching, i.e. by plunging it into a large volume of cold water. Contrary to our experience with light alloys, it is impossible to preserve the high temperature structure as a structure 'frozen in' at room temperature: however quickly the cooling takes place, the γ crystals are transformed into the α form. Nevertheless, the rapid quench does prevent large displacements of the carbon atoms and these remain dispersed uniformly throughout the newly formed α-iron.

Several aspects of this need to be specified in more detail. The transformation from γ-iron to α-iron is not necessarily complete. The steel

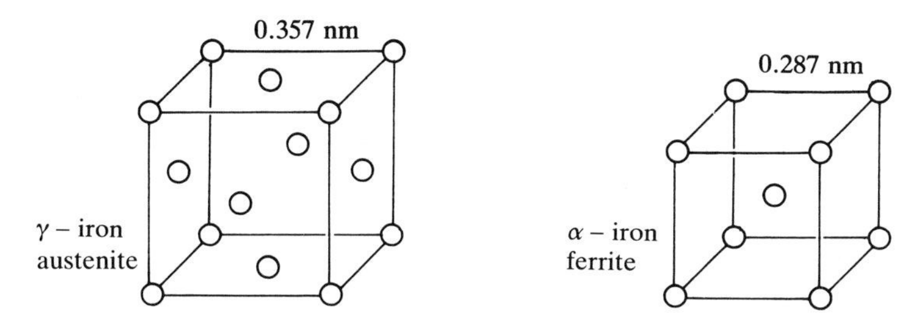

Fig. 4.26. Polymorphism in iron.

can be quenched from 950°C to a temperature different from room temperature, and the proportion of austenite still remaining depends on this final temperature. Thus, for the alloy in question, the transformation begins at 320°C and is almost complete at 110°C. The transformed part appears in the form of fine needles, whose directions are not random but are related to the symmetry axes of the γ crystal in which they are inserted. The orientations of the α grains are therefore related to the principal axes of the γ lattice. The crystallographic transformation takes place by abrupt movements of groups of atoms (it is even possible to record the noise produced when these movements are triggered).

The phase formed in the quench, known as **martensite**, consists of α-iron crystals with a greatly distorted lattice. There are two reasons for this: firstly, because of the very abrupt change in conditions accompanying its formation, the martensite contains a large number of defects and, in addition, the needles distort the austenite surrounding them. Secondly, since it contains all the carbon which was dissolved in the austenite and which is not very soluble in the α-iron at equilibrium, the martensite is supersaturated with the excess carbon and this causes considerable distortion of the α-iron lattice.

It is these large perturbations of the structure that are responsible for the exceptional strength of the quenched steel. The movement of dislocations is blocked by the martensite needles, and is impeded in the distorted regions of the austenite. After quenching, the 0.5 per cent carbon steel has an elastic limit of 1000 MPa, a very high value compared with any we have quoted so far. Very little plastic deformation occurs in quenched steel: any increase in the external stress causes the metal to break. A steel of this type is thus brittle and has very little ductility, so that, in spite of its great hardness, it is almost unusable because of the risk of failure.

Because of this, attempts are made to improve its ductility at the expense of a modest reduction in hardness. This is the aim of the process known as **tempering**, achieved by **annealing**. After quenching, the steel is reheated to a temperature below 800°C and certainly well below that at which any of the original austenite might be re-formed. The martensite first becomes heterogeneous, with regions where carbon atoms assemble in groups. After that, when the annealing temperature rises above 250°C, the mobility of the atoms is sufficient for the formation of a new crystalline phase: Fe_3C, or cementite. The steel then contains two phases: **ferrite** (i.e. the α-iron without carbon) and **cementite**. The ferrite grains are very small, but the cementite grains are even smaller and are dispersed amongst the ferrite. These two factors, the small grain size and the dispersion of the cementite, give the metal strength without the loss of all its ductility, so that it is no longer brittle. By choosing appropriate annealing conditions (temperature and time), the metallurgist can, within certain limits, vary the properties of steel in any required direction.

The procedures for quenching and annealing of steel were for a long time a matter of craftsmen's recipes, but theory has now justified them and enabled them to be taken further. Two recent examples of advances in the iron and steel industry are of considerable importance: **maraging** and **ausforming** steels.

Maraging steel

Consider an alloy of iron and nickel (18 per cent) with additions of cobalt (6 per cent) and molybdenum (5 per cent), forming a steel containing practically no carbon. As with other steels, this one is austenitic at high temperatures, where the added elements are soluble, and is martensitic when cooled. This is an alloy with some very special properties: it does not need quenching for the martensitic transformtion to be induced — it is sufficient to allow an ingot to cool in the free atmosphere. The advantage of this is that the martensitic transformation is then not just an effect that can be induced to occur in objects thin enough for a quench to be effective, but it can be made to occur throughout the body of a component, even one with a large volume.

Other features of the nickel martensite are:

1. it is not only very strong but is not brittle;
2. it can be reheated up to 500° C without such an anneal causing it to decompose.

During this treatment, however, the added elements, which are in a supersaturated state, are precipitated in the form of very finely dispersed grains. The structural hardening resulting from this fine precipitate is thus added to the hardening process due to the formation of martensite. Very high elastic limits for steels (2000 MPa) can be achieved in this way without sacrificing any of their ductility.

Thermomechanical treatments

These involve a combination of the effects of heat and deformation. In **ausforming**, for example, the steel in an austenitic phase, and therefore at high temperature, is subjected to a considerable deformation, e.g. to very heavy rolling. It is then quenched to produce the martensite and finally tempered by annealing. The violent mechanical treatment at high temperature produces extremely small grains, and these remain very small during the subsequent quench and anneal. This so improves the quality of the steel that elastic limits of 3000 MPa can be attained. The importance of this technique lies in its ability to give an ordinary low-alloy steel qualities that are comparable with those of much more expensive special steels.

Interaction of dislocations with impurities: ageing effects

A metal containing a small concentration of foreign atoms could be regarded as a very dilute alloy, but it is generally considered more as a metal with impurities since the extra atoms have often not been introduced deliberately: they come either from the raw materials or from contamination during the processes involved in the production of the metal.

The impurity atoms are randomly distributed and produce distortion in their immediate vicinity, particularly if they occupy interstitial sites in the crystal lattice. A lattice is disturbed along a dislocation line and in its immediate neighbourhood, and here the interstitial sites between sets of neighbouring atoms have a larger volume than they have in the perfect crystal. If an impurity atom moves near a dislocation during its migration within the crystal due to thermal agitation, it finds sites that will accommodate it so easily that there is little chance of it moving away again. The dislocation becomes 'decorated' in this way with a cloud or atmosphere of foreign atoms which is produced in a time that is significantly greater than the duration of experiments (varying from a few seconds to a few tens of minutes, depending on the temperature).

The impurity cloud that condenses along the line of a dislocation holds it more stable in position, so that a greater stress is needed to move it than would be necessary if the dislocation were 'bare'. If the dislocation is freed by a large enough stress, it is propagated too quickly for the impurity cloud to follow it. When the dislocation stops, however, the cloud re-forms and, after a certain time, it once again pins the dislocation in place.

A model of this type accounts for several aspects of the ductility of certain metals which at first sight appear to be very strange.

The upper yield point on the tensile curve for mild steel

Figure 4.27 shows the tensile curve given by an extra mild steel (less than 0.1 per cent carbon) used in car-body panels. The steel may contain nitrogen as an impurity, apart from the carbon. When subjected to stress, the elastic region terminates at a limit known as the upper yield point where the stress is R_{es}. As soon as yielding starts, however, the stress falls (the **yield drop**) to a lower value R_{ei} at the lower yield point, where the

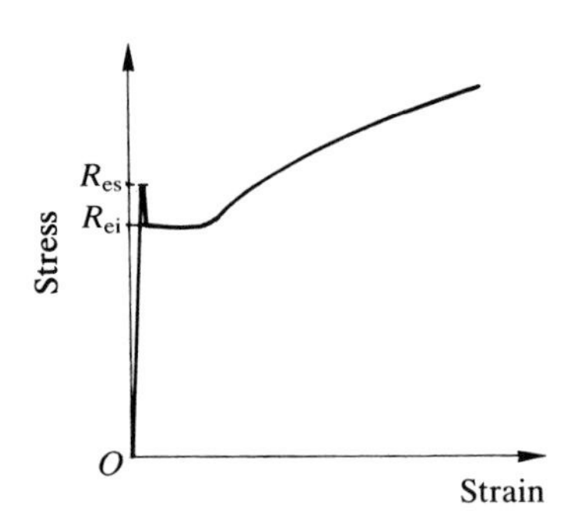

Fig. 4.27. The upper yield point on the tensile curve of mild steel.

extension is now several per cent. After that, the usual stress–strain curve is followed, during which the normal processes of plastic deformation and work-hardening occur.

In the metal prior to deformation, the dislocations are pinned by their impurity clouds (of C and/or N atoms). When the stress reaches the value R_{es}, the dislocations are set in motion and are thus freed from their clouds. These bare dislocations, as well as those created by active sources, which also lack impurity clouds, can all be moved by the lower stress R_{ei}. The two values for the elastic limit both correspond to dislocations being set in motion, but in one case they are accompanied by their impurity clouds and in the other they are 'bare'.

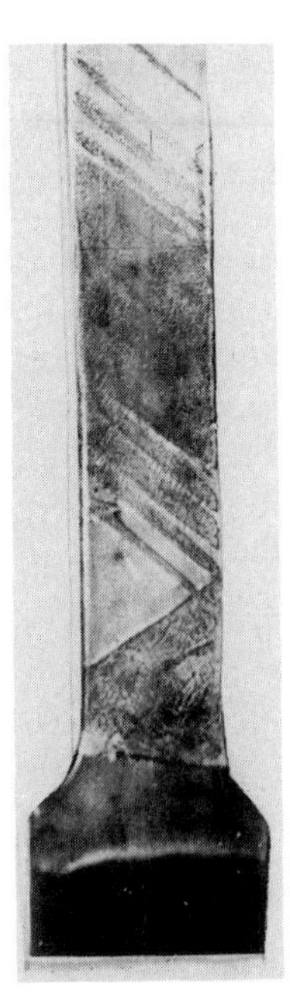

Fig. 4.28. Lüders bands in a mild steel under tension at room temperature.

At this stage (stress R_{ei}), the deformation is not uniform but is concentrated in bands (Lüders bands) across the test specimen, as can be seen in Fig. 4.28. It is not until the specimen is completely covered by these bands that the plastic deformation continues as usual under a continually increasing stress.

Ageing

There is one type of experimental observation that very clearly reveals the effect of impurities in mild steel: this is the phenomenon of ageing. At the stage when work-hardening has occurred (extension ϵ_1, Fig. 4.29), suppose we stop any further deformation by removing the stress σ_1. After a very short time, the stress is applied once again. The metal behaves normally: it deforms elastically up to the stress σ_1, and the plastic deformation continues as if there had been no interruption. However, if the time during which the stress is removed is long enough (at least an hour), the upper yield point that was observed with the fresh mild steel specimen occurs once again. This shows that the impurity cloud has had enough time to re-form around the dislocation so that, after ageing, the dislocations have to be freed from their clouds once more, whereas they were still bare when the strain was almost immediately reapplied.

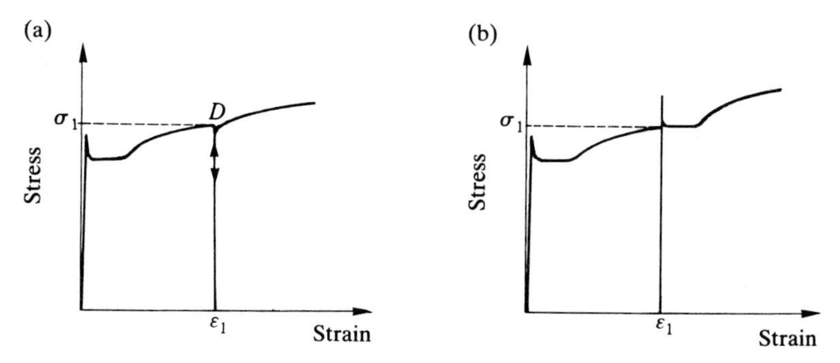

Fig. 4.29. Ageing tensile curves for a reloaded test specimen. (a) Reloading immediately after unloading at D; (b) reloading after ageing for one day.

Blue brittleness

Craftsmen have long known that mild steel becomes brittle at a temperature near 150°C, where treatment in air covers it with a thin oxide layer and gives it a blue colour. The impurity clouds on the dislocations form much more rapidly at these temperatures than at room temperature, since the diffusion of impurity atoms through the metal is much easier. As soon as the dislocations are released on reaching the stress R_{es}, they are joined almost at once by their impurity clouds: the strain at the stress R_{ei} cannot occur as it did in Fig. 4.27. Even for small extensions, the plastic stress exceeds R_{es} and is observed to increase rather irregularly, so that for an extension that is still quite small the failure stress is reached. The steel behaves almost like a brittle material.

Other modes of plastic deformation

A very general feature of plastic deformation is its non-uniformity, arising from the fact that the crystal defects which play an essential role in its initiation are themselves not distributed uniformly throughout a crystal. Slip, for example, is always concentrated in a series of fine bands. Although the numerous slip bands that normally occur are distributed fairly regularly over a specimen, there are exceptions, as we have already seen in the case of mild steel (Lüders bands, Fig. 4.28).

We give a couple of other examples below: true, these are special cases, but they do show quite dramatically the complicated and sometimes unpredictable behaviour of a real solid. We must be wary of imagining that we know everything about a metal when we have a complete knowledge of its crystal structure: it is the real structure, with all its defects, that we need to know, even if it departs only very little from the ideal structure.

1. Consider a rod formed from a single crystal of zinc and subjected to compression along its axis. The most uniform plastic deformation that could possibly occur would be a shortening of the rod with an accompanying increase in its diameter. However, what can happen is that the shortening is the result of a very localized bend in the specimen, like the 'kink' shown in Fig. 4.30, with the rest of it remaining virtually intact. Zinc

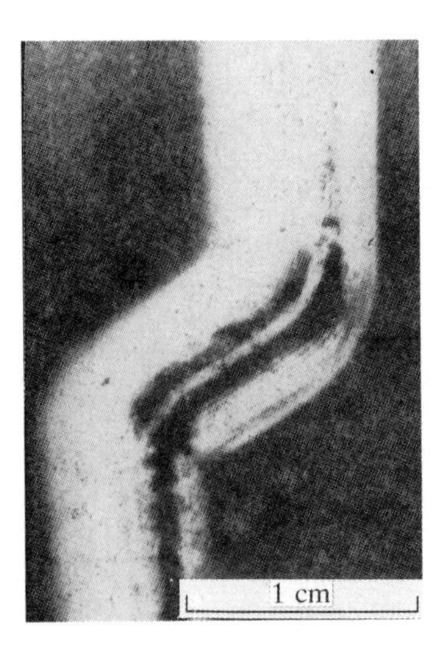

Fig. 4.30. A kink in a zinc crystal (by kind permission of C. Barrett).

crystallizes in the hexagonal system and this type of crystal has fewer possible slip systems than a cubic crystal. It can be appreciated that there will be some orientations in which a uniform strain cannot be achieved by a combination of slip systems. Models of the dislocation distributions that might explain the bending of the crystal have been devised, but it should be realized that we do not know why the bending takes place nor where it might occur in a given specimen.

2. **Twinning** is described in texts (e.g. SM, p. 118) on the structures of real crystals formed by the stacking of hexagonal planes of atoms. The two parts of the twin have the same intrinsic crystal structure, but are arranged in such a way that one is the mirror image of the other in a plane located in the twin plane. To change a single crystal into a twinned crystal, the atoms need to be displaced by a shear stress applied parallel to the twin plane. If a given deformation of a specimen (e.g. its extension by a tensile stress) happens to produce changes in shape that coincide with those required to produce a twin, then it is clear that twinning could be initiated by external forces (mechanical twinning).

In some metals, the existence of twins is revealed by examining the polished surface of a strained specimen under the microscope. The pairs of crystals forming the twins are recognizable because they are separated by a perfectly straight junction (Fig. 4.31). The production of a twin by the displacement of atomic planes in a small volume is abrupt and is sometimes audible. For example, if a rod of tin is twisted by hand, a series of clicks or creaks can be distinctly heard (the 'cry of tin'), each corresponding to the appearance of a set of small twins in the tin crystals.

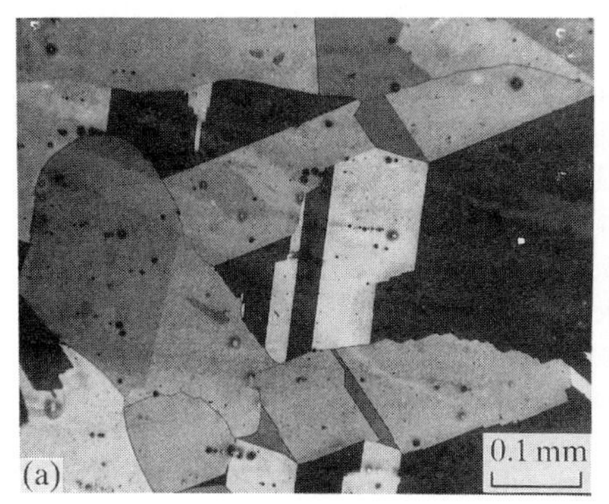

Fig. 4.31. Twinned crystals: (a) under an optical microscope (by kind permission of P. Lacombe); (b) under an electron microscope (by kind permission of C. Donadille).

Plastic deformation by twinning occurs mainly in crystals where only a few slip systems are possible: metals with hcp (hexagonal close-packed) lattices twin more easily than those with a cubic lattice. There are some orientations of a crystal with respect to the applied stress for which deformation is more easily accomplished by twinning than by slip.

Once again, we have to rely on qualitative explanations: we do not know the precise conditions under which a twin can be produced in a given crystal subjected to a given external stress.

Fracture

It is vital that no component of a machine or structural element of a building should fail in service. This is clearly an absolutely essential condition from the point of view of safety, but it is also a condition for the durability of an installation, which is an important economic consideration. To prevent accidents, engineers have to calculate the stresses at every point in a given machine or structure in the light of the operational requirements placed on it, and this is what theoretical mechanics enables them to do. They also need to know the intrinsic properties of the materials, and these are provided by the systematic tests carried out on the products of the metallurgical industry.

Brittle fracture

We start with a simple case, which is also one of the most typical: that of a glass or a crystal of rock salt. When solids like these are subjected to a gradually increasing tensile stress, they suffer only very small elastic extensions before breaking suddenly and without any plastic deformation. This is known as **brittle fracture**.

In a glass rod, the break occurs in a direction nearly normal to the direction of the tension: the surface that is revealed is rounded and smooth. In crystals, the break is parallel to a lattice plane and they are said to cleave. In a specimen of NaCl, for example, the cleavage occurs in a plane parallel to one face of the cubic unit cell. If the solid is formed from a polycrystalline mass, as is generally the case for crystalline solids, the break reveals facets that are oriented in various directions since the cleavage planes have different orientations in the different grains (Fig. 4.32). In a polycrystalline metal, brittle fracture can also take place through a loss of cohesion between adjacent grains.

The cleavage of a NaCl crystal can be very simply described on an atomic scale (Fig. 4.33). The Na^+ and Cl^- ions are arranged in dense planes of atoms that are identical: cleavage separates two adjacent atomic planes P_1

Fig. 4.32. Brittle fracture: (a) fracture of iron by cleavage at low temperature; (b) intergranular impact fracture (scanning electron micrograph, by kind permission of G. Henry).

and P_2 a distance a_0 apart. In equilibrium, the force between these two planes is zero: the attractive Coulomb force between ions of opposite charge is exactly balanced by the repulsive force between the electron shells which cannot penetrate each other. When the distance between the

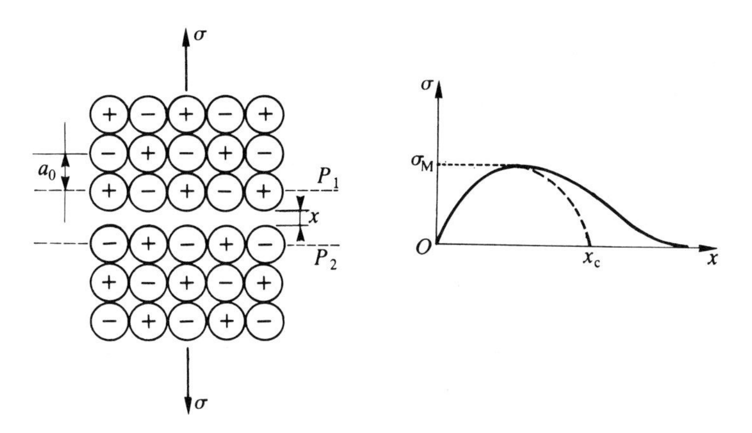

Fig. 4.33. Calculation of the theoretical stress needed for the cleavage of NaCl.

planes is greater than a_0, the attractive force becomes predominant, passes through a maximum, and then decreases to zero as the planes move far apart. Under the action of an external stress, the planes P_1 and P_2 are pulled apart, although only through a relatively small distance since we are in the elastic region. When the stress exceeds the maximum value σ_M, equilibrium can no longer be maintained and cleavage occurs.

This atomic model is simple enough for us to be able to use it in making a theoretical estimate of the stress σ_M needed for fracture by cleavage.

Box 20

Calculation of the theoretical stress needed for the cleavage of rock salt

A stress σ is applied to a crystal of unit cross-section in order to separate two adjacent atomic planes by a distance $a_0 + x$, where a_0 is the equilibrium distance between the planes. At the beginning of the deformation, when the strain is elastic, σ is proportional to x; it then passes through a maximum value σ_M, decreases and eventually becomes zero: the planes are then completely separated.

We represent σ approximately by the sinusoidal variation $\sigma_M \sin(\pi x/x_c)$ up to x_c and by zero if $x > x_c$ (Fig. 4.33).

The strain is small in the elastic region, so that the sine can be replaced by its argument: the stress is then $\pi\sigma_M x/x_c$. Since Young's modulus E is defined by $\sigma = Ex/a_0$, we can relate E to σ_M/x_c:

$$\frac{\sigma_M}{x_c} = \frac{E}{\pi a_0}.$$

When the crystal is split into two pieces, two free surfaces are created and this increases the energy of the system by 2γ, where γ is the surface energy of NaCl. We identify this increase with the work done by the tensile stress in separating the planes, i.e. with the integral:

$$\int_0^{x_c} \sigma_M \sin\frac{\pi x}{x_c}\, dx = \frac{2\sigma_M x_c}{\pi}.$$

We thus obtain the relationship:

$$2\gamma = \frac{2\sigma_M x_c}{\pi}$$

or

$$\sigma_M x_c = \pi\gamma.$$

Using the previous result that:

$$\frac{\sigma_M}{x_c} = \frac{E}{\pi a_o},$$

we arrive at:

$$\sigma_M = \sqrt{\frac{E\gamma}{a_0}}.$$

It turns out that the experimental values for the stresses needed to fracture any crystal are much less than the theoretical predictions. This is the same situation as occurred with the elastic limit of ductile solids, i.e. a considerable discrepancy between calculation and experiment. We are led to a similar conclusion: the model of the perfect crystal is inadequate for an understanding of fracture phenomena. The effective strength of a brittle crystal is in fact limited by defects in the structure.

There was also something a little illogical in our model of cleavage. In a perfect crystal, all the planes are identical: there is no reason why cleavage should occur between two planes P_1 and P_2 rather than any other two. If these two planes are distinctive in some way, it can only be because some irregularity or defect is present in their locality.

While the defects responsible for plastic deformation are on an atomic scale, those connected with fracture are more extensive: of the order of a micrometre rather than a tenth of a nanometre. In 1920, Griffith put forward an idea which still forms the basis of theories of brittle fracture, an

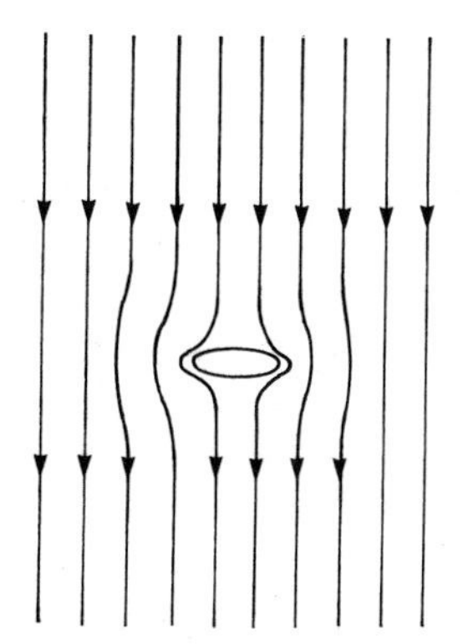

Fig. 4.34. Concentration of stress lines at the ends of an elliptical cavity.

idea that is suggested by everyday experience. If we have two strips of paper, one with perfectly smooth edges and the other with edges nicked by a small tear, and try to pull each of them apart with a sudden application of the same tensile stress, the first offers strong resistance while the second splits easily into two. The small tear in the edge has initiated the fracture of the strip.

Now consider a sheet of uniform material which has a transverse crack in the surface or an internal cavity with a width much smaller than its length. When a tensile stress is applied to the sheet in a direction perpendicular to the crack or to the longer dimension of the cavity, the lines of force of the stress field circumvent the crack or cavity and become much denser around their extremities (Fig. 4.34). A concentration of the stress thus occurs at the ends of the crack or cavity, which can be calculated if it assumed that the whole specimen is strained elastically. Some idea of the order of magnitude of the stress concentration produced can be obtained by taking the case of a hole having an elliptical cross-section with an axial ratio of 10. The maximum stress in the surface around the point at the end of the cavity in this case is 20 times the applied stress. In regions like this, therefore, the effective stress experienced by the material may exceed the theoretical stress needed for cleavage, which can then be initiated at a defect. The specimen may fracture, even though the overall stress applied to it is quite small.

Microcracks occur over the surface of any crystal, which then breaks at the point where the stress concentration is greatest. Given the irregularity of natural microcracks, fracture can be expected to occur for stresses which are always very much lower than the theoretical values predicted for perfect crystals, but which vary quite a lot from one specimen to another. This is confirmed by experiment.

Several observations support the Griffith theory. If the surface of a rock salt crystal is dissolved away, any superficial cracks that might have existed on the specimen are removed. If it is then subjected to a tensile test immediately after this treatment, the stress required to fracture it is found to be higher. Certain experiments carried out with glass are also quite convincing. A thin glass rod freshly drawn in a flame is stronger because its surface is very smooth and defects have not had the chance to occur. It is, however, very easy to make them appear: a single contact with a steel needle is enough to cause a small scratch which, even if it is no more than one or two micrometres deep, reduces the fracture strength. If the fibre is elastically bent, a scratch is only effective when it occurs on the surface in tension, where it can open up and be propagated: when it occurs on the surface under compression, it closes and is blocked.

This effect is the basis of a process for reducing the brittleness of glass sheets ('toughening'). By cooling the surfaces quickly (a superficial quench), they are put under compression, thus reducing the embrittling effect of accidental surface scratches.[†] A similar application is the method used to increase the toughness of ceramics by glazing. It is impossible to reduce the number of surface cracks to any appreciable extent and attempts are therefore made to increase the resistance to their propagation instead.

Brittle fracture in metals is more complex than that in glasses or ionic crystals. It is well known that metals are not breakable in the same way as glass. There is always a certain amount of ductility but, when they *are* brittle, the amount of plastic deformation before fracture is very small.

In this case as well, it is thought that failure is the result of the propagation of microcracks or microcavities. However, except in fatigue fracture (p. 234), the centres where the fracture is initiated are generally internal and no direct experimental data on them are available.

Several mechanisms have been suggested for the formation of these internal defects. In a dislocation, there is a line of voids at the edge of the interrupted plane of atoms: if several dislocations glide in the same plane and are blocked by an obstacle such as a grain boundary, they coalesce and form a microcavity which can grow further by the addition of vacancies or the loss of cohesion between the side walls of the cavity (Fig. 4.35). Microcracks can also be produced at the surface of grains of a second phase. Thus, graphite flakes in grey iron (SM, p. 208) make it brittle because of a loss of cohesion between the graphite and the ferrite. The loss of cohesion may be due to the segregation of impurity atoms along the grain boundaries, which modifies the interatomic bonding at the grain surface. This can be described macroscopically as a reduction in the surface tension of the metal: thus, the free surfaces of the cavity can grow quite easily since only a small input of energy is required.

In metals that are brittle but nevertheless retain a certain ductility, the atoms in the surface of microcavities or cracks have a tendency to rearrange themselves so as to round off the sharp angles between the walls. As a result, the increase in stress is lower for a metal than for brittle

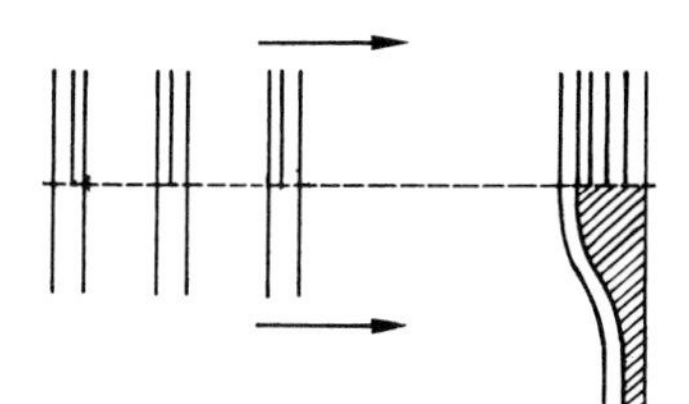

Fig. 4.35. Formation of a microcrack by the coalescence of blocked dislocations at a grain boundary.

[†]Toughened glass produced in this way (e.g. car windscreens) can be recognized by interference colours due to double refraction arising from internal stresses.

non-metallic materials like glasses and the discrepancy between real fracture strengths and theoretical values is therefore less.

A crack initiated by a microcavity in a polycrystalline metal must, if it is to grow, propagate itself from grain to grain. However, because of the different orientations of the cleavage planes in adjacent crystals, the crack is deviated after crossing the boundary. This hinders its extension, at least at low temperatures when thermal agitation is limited. The smaller the grain size, the more obstacles there are to be surmounted in a given length of path and the greater the stress required to cause fracture. Any treatment which produces a smaller grain size is useful in reducing the brittleness of a metal (as it is in increasing the elastic limit (p. 206)).

Brittleness seriously handicaps the technologist: it prevents the use of very hard materials (e.g. fibres of carbon, glass, silicon carbide, etc.) which are strong but liable to premature fracture. A solution to the problem is provided by composite materials in which the fibre is embedded in a plastic matrix that opposes the propagation of any microcracks existing in the surface. This removes the centres from which fracture is initiated and the brittleness is greatly reduced. Some present-day composites are among the strongest materials ever produced and their use in such sports activities as polevaulting and in forms of transport from sailing boats to spacecraft has brought the attention of the advances in this field to a very wide public.

Ductile fracture

When metals are taken beyond their elastic limit, they generally undergo plastic deformation. In brittle materials this process is very quickly interrupted, but even when they are not brittle the deformation does not go on indefinitely but ends in a type of fracture known as **ductile fracture**.

When a specimen is plastically deformed in a tensile test, for example, the extension is not uniform but is concentrated in a short region where the cross-section becomes smaller: this is known as **necking** (Fig. 4.36 — see also p. 191). Once it starts, the necking can only increase since the reduced cross-section causes the stress in the narrowed region to rise.

If the metal were perfectly plastic, the specimen would become increasingly narrow at the neck until the cross-section was zero. In reality, it breaks before that: fracture occurs at the neck when the diameter has decreased by an amount that varies with the type of metal and even from one specimen to another.

The appearance of the fracture surface is not the same as in the case of brittle failure. There are no facets corresponding to the cleavage of grains, but an irregular surface with quite a soft or fibrous profile (Fig. 4.36). In this part of the metal, the grains have been broken and the metal is very greatly deformed.

Fig. 4.36. (a) Ductile fracture with necking; (b) appearance of the ductile fracture surface of a stainless steel (electron micrograph, IRSID).

In this case as well, fracture is due to the propagation of a crack across the specimen. It is initiated by the growth of microcavities produced during the large deformation of the metal at the neck at sensitive points such as grain boundaries, small grains of a more brittle phase, etc.

The transition between ductile and brittle fracture

Many materials can be classified unambiguously in one of the two categories described above. However, there are some which may be ductile or brittle depending on the conditions: the temperature, the rate at which stresses are applied, the purity of the metal, etc. Mild steels (i.e. those containing less than 0.5 per cent carbon), together with other metals having a body-centred cubic lattice, become brittle below quite a well-defined transition temperature.

Brittleness is measured by the **Charpy V-notch impact test** (Fig. 4.37). The specimen is in the form of a bar with an accurately defined shape and has a groove cut in it in order to facilitate and localize the fracture. The ram is dropped from a known initial height and, after its sudden impact on the specimen, its remaining kinetic energy is sufficient to carry it through to a second measured height. The energy absorbed by the fracture is calculated from the two heights: the more brittle the material, the less the energy absorbed by fracturing.

The ductile–brittle transition occurs over a temperature range of about ten degrees. The transition temperature varies with the type of steel: for many, it occurs a little below 0°C, i.e. within the range of possible ambient temperatures. This is what gives the effect its practical importance. An

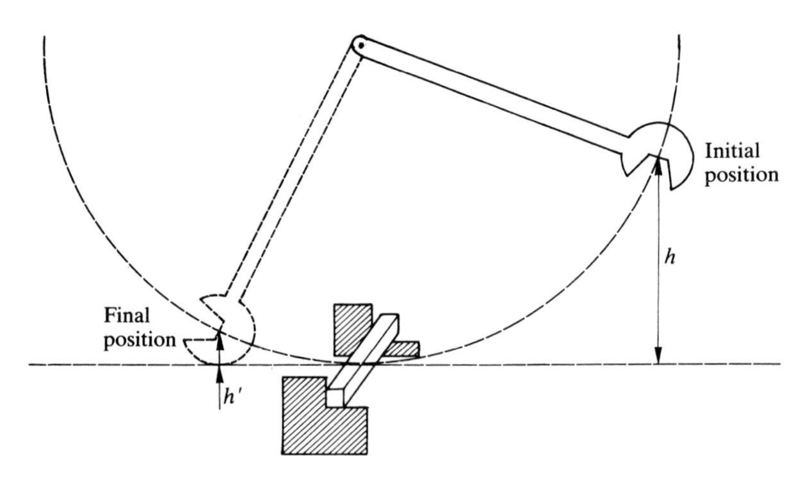

Fig. 4.37. Measurement of impact strength by the Charpy V-notch test. The ram is released at a height h, and carries on to the height h' after breaking the specimen. The energy absorbed by the impact is $mg(h - h')$.

example that illustrates the point very well is that of the barrels of the heavy artillery which broke up during the severe winter of 1939–40 (although this did not surprise the gunners, who were often of peasant origin and were very familiar with tools that fractured during very cold weather). Another example is the break-up of *Liberty ships* in service.

The existence of a brittle transition is explained as follows: the elastic limit of mild steels is greatly increased at low temperatures. This is partly because dislocation sources become less active but mostly because the mobility of the dislocations falls: this arises from the particular structure of the dislocation core in body-centred cubic crystals. At low temperatures, therefore, plastic deformation only starts when the stress is very high. On the other hand, the stress required to initiate crack propagation does not vary much with temperature. Thus, below the transition temperature, a stress that is enough to initiate the propagation of cracks is reached before plastic deformation has begun: the metal then behaves like a brittle material. Above the transition temperature, on the other hand, the plastic deformation is already appreciable when the applied stress is still insufficient to activate the centres initiating fracture.

The embrittlement of a metal is a complex phenomenon because temperature is not the only factor involved. The conditions under which brittleness tests are carried out on metals must therefore be very precisely specified. We shall discuss in turn the method of distributing the stresses in the specimen, the rate at which the stress is applied and the presence of impurities.

For body-centred cubic metals at room temperature, a notched specimen is brittle whereas one with a smooth surface is not. The complex system of stresses distributed around the notch can block any plastic deformation, so that the fracture load may be reached without any deformation having occurred.

If the stress is applied to the metal very abruptly by an impact, the specimen may break, whereas it would have withstood the same stress if it had been applied slowly and steadily. We have already mentioned the fact that the rearrangement of atoms in the surface of microcracks, by making the angles less acute, reduces the stress concentration. During an impact, such rearrangement does not have time to occur and so, around the initial crack, the stresses are high enough for fracture to begin.

It seems surprising that very small changes in the composition of a metal should have significant effects on its brittleness, yet it is observed that the addition of a few parts per million of an element can induce brittleness and the complete removal of a very minor impurity can eliminate it. The explanation of this is that only a few small and scattered grains of a second phase are needed to create centres at which fracture can begin. Moreover, when impurity atoms collect at grain boundaries, their surface concentration may be large enough to have an effect on intergranular cohesion, even though their overall density seems negligible.

We can give several examples of this: chromium was long considered a brittle metal, but this was simply because there was no known method of preparing it with a sufficient degree of purity; the compound Ni_3Al, which is involved in heat resistant 'superalloys', loses its brittleness with the addition of a trace of boron; gallium, on the other hand, can be regarded as a 'poison' as far as aluminium is concerned since it drastically reduces intergranular cohesion.

Creep

So far, we have implicitly assumed that the deformation of a solid depends on the stress applied to it at a given temperature. In fact, it is true that a permanent equilibrium state is reached, at least at low temperatures: 'low' meaning, say, a temperature below half the melting point expressed in K.

At high temperatures, however, the plastic deformation under constant stress increases with time and therefore depends both on the applied stress and the time of application. This phenomenon, known as **creep**, becomes significant (and complicates our interpretation of observations) not only when a metal is deformed slowly but even during rapid deformation at a high enough temperature, e.g. as in the thermomechanical treatments mentioned above (p. 214).

Creep can be measured using a tensile testing machine in which the specimen is placed in a vessel at a constant but adjustable temperature. A

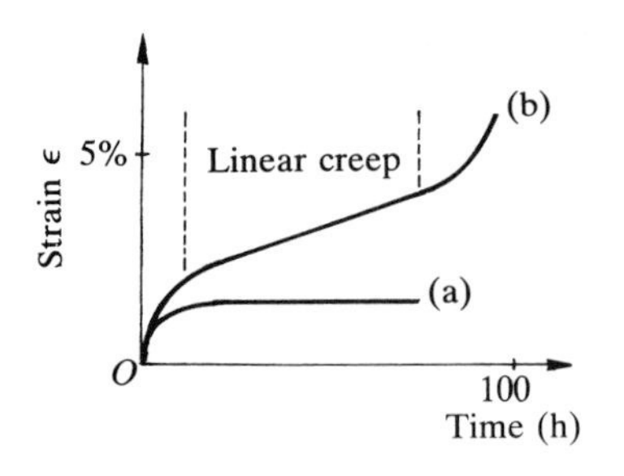

Fig. 4.38. Creep curves for the same stress: (a) at low temperature ($<\frac{1}{2}T_F$); (b) at high temperature.

constant stress greater than its elastic limit is applied to the specimen and its length is then plotted as a function of time (Fig. 4.38). At low temperatures, there is an initial almost instantaneous deformation, after which the rate of extension steadily decreases until a limiting length is eventually reached. At higher temperatures, on the other hand, after the initial transient stage, the extension continues at a uniform rate and there is then a third stage during which the extension accelerates until the specimen finally breaks.

In the 'linear' stage, the rate of extension with time increases with the temperature and the applied stress. Creep is only significant if the temperature is higher than about 40 per cent of the melting point expressed in K, but it then becomes appreciable: the extension can be as much as several per cent and can last for hundreds, or even thousands, of hours. The following figures for a widely-used steel containing 0.4 per cent carbon, 0.24 per cent silicon, and 0.9 per cent manganese give some idea of the orders of magnitude involved: with the steel subjected to a stress of 7MPa, the creep is 2 per cent after 5 hours at 670 °C, after 50 hours at 625 °C, or after 500 hours at 585 °C.

It is obviously essential that no appreciable creep should occur in a machine in permanent use, and this clearly limits the operating temperature of a metal. The effect assumes the greatest importance in machines which, by their very nature, have to operate at the highest possible temperatures (e.g. the rotors in jet engines).

The atomic mechanisms involved in creep

In looking for mechanisms on the atomic scale to account for creep, we shall confine our attention to the second linear stage. The tertiary stage, where the effect accelerates, simply corresponds to the onset of damage in the metal: cracks and cavities are formed, develop, and become centres for the initiation of fracture.

As is the case during the plastic deformation accompanying cold working, creep in metals involves the multiplication of dislocations by active sources and their glide along lattice planes within the crystal grains.

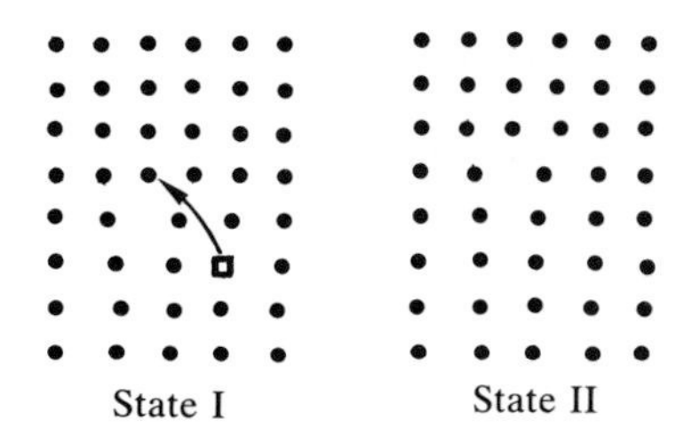

Fig. 4.39. Dislocation climb produced by vacancies diffusing through the lattice.

However, other types of atomic motion can occur because the higher temperature increases atomic mobility so that **diffusion**, which is discussed in Chapter 5, is then possible. For example, atoms diffusing through the crystal may arrive at a dislocation line or may diffuse away from it. One result of this is that the dislocation, instead of moving parallel to the slip plane (glide), is moved in a direction perpendicular to it (**dislocation climb**, Fig. 4.39).

There are also deformation processes not involving dislocations. For example, a crystallite under tension becomes elongated if atoms in the lateral surfaces of the grain migrate towards the end faces by diffusion (Fig. 4.40). It has also been demonstrated experimentally that two adjacent grains in a polycrystalline metal can slip with respect to each other, while remaining contiguous, because of the atomic disorder existing along the boundary.

During low temperature plastic deformation, dislocations are increasingly multiplying and modifying the state of the metal: this is the work-hardening we have already discussed. In high temperature creep, on the other hand, the metal reaches a steady state: the disorder due to the deformation is compensated by a restoration of order through local

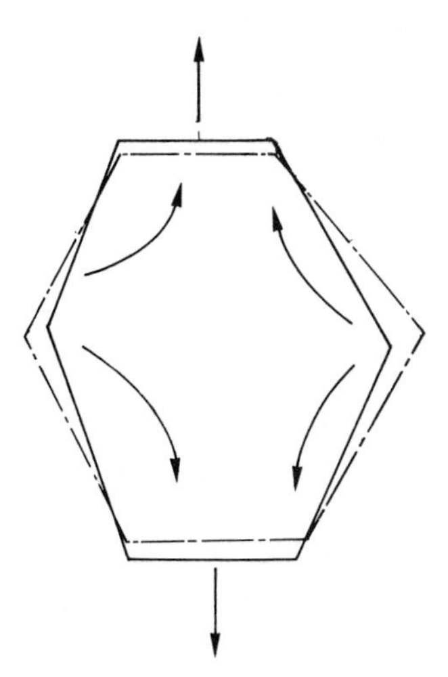

Fig. 4.40. Elongation of a grain by the migration of atoms from the lateral boundaries to the upper and lower ends.

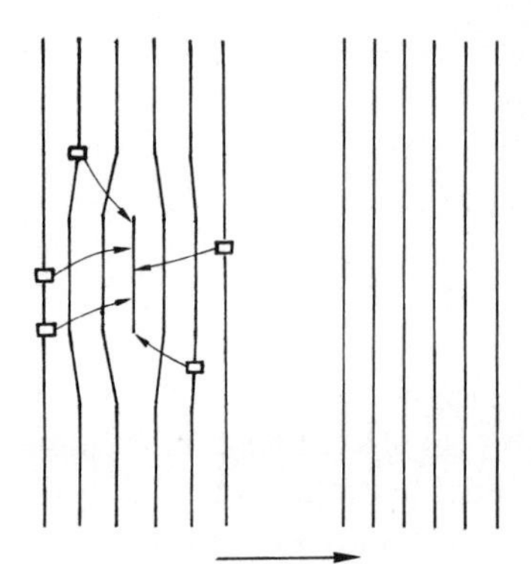

Fig. 4.41. The annihilation of two dislocations of opposite signs: the atoms become dispersed in lattice vacancies.

rearrangements of atoms by diffusion. Pairs of dislocations of opposite signs annihilate each other (Fig. 4.41), and dislocations gather in such a way that a rather distorted domain is replaced by two more regular domains with different crystallographic orientations separated by a sub-boundary where the dislocations are assembled. This is an effect known as **polygonization** (SM, p. 116).

The various mechanisms we have been describing involve the displacement of atoms from one lattice site to another by diffusion (Chapter 5). This is a slow process, unlike dislocation glide which occurs at a speed close to that of sound, but diffusion rates do increase very rapidly with rising temperature. This accounts for the main features of creep: a slow process that is only significant at high temperatures.

Solid state physicists have been able to quantify all the qualitative ideas mentioned above, and their calculations have shown that the various mechanisms envisaged are possible and have effects conforming to macroscopic observation. One thing is certain, however: creep is a very complex phenomenon. We may understand why it occurs, we may be able to account for the orders of magnitude of the observed effects, but we do not yet have an indisputable quantitative picture of it.

High-temperature alloys

When the performance of a machine improves as its operating temperature increases, it is ultimately limited by the need to remain in a temperature range where no significant creep occurs in the main components. The jet engines powering aircraft are a typical example. How can we produce creep-resistant materials for use in such cases?

One solution would be to use materials with very high melting points: refractory materials like tungsten (MP = 3663 K) or molybdenum (2883 K). Since creep is negligible at temperatures below 40 per cent of the melting point, operating temperatures above 1100°C could be achieved and this would be very worthwhile. Unfortunately, refractory metals are too readily oxidized in air to be usable.

The best materials available at present are alloys, known as **superalloys**, based on nickel and chromium with aluminium and titanium as additional elements. Their first desirable property is that, when exposed to air, they are covered by a continuous coating of chromic oxide which protects the surface and prevents oxygen from penetrating the alloy: this is the same effect that is used in stainless steels. The second and more important feature of these alloys is that they can be used at temperatures considerably above 40 per cent of their melting points (which are around 1350°C) without appreciable creep. Such unusual creep resistance is due to the presence within the nickel–chromium matrix of precipitates containing the additional elements in the form of very finely dispersed grains. These provide obstacles to the movement of dislocations and thus oppose any deformation. Although this is the same process as that occurring in hardened light alloys (p. 211), the latter only harden at low temperatures: as soon as the temperature rises above about 250 to 300°C, the precipitates coalesce into large grains which are too widely spaced to retain any hardening effect. In superalloys, the precipitates remain finely dispersed up to 1200°C. A possible explanation for this is as follows: precipitates with the composition Ni_3Al crystallize with a lattice very similar to that of the nickel matrix. These form obstacles sufficient to block the movement of dislocations, but the lattices are coherent enough for any atomic disorder at the interfaces to be minimal. Because of this, the surface energy of the grains included in the matrix is small, so that their enlargement would not lower the energy of the system very much through the reduction in the interfacial area: such enlargement does not therefore occur. Above 1050°C, however, the solubility of the added elements increases: the precipitates that are responsible for the creep resistance progressively disappear as the temperature rises and the performance of the alloy deteriorates.

To avoid this problem, the hardening particles should be strictly insoluble in the metal, as is the case with thorium or aluminium oxide. These can be incorporated into the metal using powder metallurgy: a fine powder of the metal is mixed with a little oxide powder (10 per cent) and the mixture is agglomerated into a solid by sintering (compression at high temperature).

This is the process used for the production of tungsten filaments for incandescent lamps. If the tungsten were pure, it would recrystallize when the lamp is in use and the grains would then be in the form of large crystals lying end to end along the filament: this could not retain its shape since it would creep under its own weight. When grains of thorium oxide are introduced, the tungsten crystals cannot become enlarged and no creep occurs in the filament even when it is incandescent.

Another advance in technique is based on the following fact: one cause of creep in a polycrystalline metal is the slip that takes place between two adjacent grains along the boundary separating them, an effect that does not occur in a single crystal. The conditions under which superalloys solidify have been so modified that a whole turbine blade, for example, is virtually a single crystal: in fact, it is more a mosaic of crystallites with very similar orientations. In this way, the atomic disorder along the grain boundaries is slight and is no longer a cause of creep. Another technique that has been developed is to induce crystallization with such an orientation that the grain boundaries are perpendicular to the principal tensile stress and so make no contribution to creep.

One result of the continual advances in the field of superalloys is that the maximum operating temperature of jet engines has, in a single decade, been raised by between 50 and 100 °C, a feat that has led to a considerable improvement in their efficiency.

Non-metallic creep resistant materials

Refractory oxides, such as those of aluminium, magnesium and silicon, have very high melting points, as do certain carbides (SiC) and nitrides (Si_3N_4). The interatomic forces in these compounds, whether ionic or covalent, are much greater than those in metals, so that they are all very hard solids. Small crystals of such substances, bound together in a compact mass, form **ceramics**, a class of material whose good mechanical properties are preserved without creep up to very high temperatures (more than 2000 °C). Research is currently being carried out with the aim of replacing some of the metallic components in motor vehicles with ceramic equivalents.

However, ceramics do not perform very well under tension because they are brittle: surface cracks are easily propagated when the materials are subjected to tensile stresses and they break. Under compression, on the other hand, the cracks close up and are not propagated, so that the materials can withstand considerable compressive loads even at very high temperatures. That is why ceramics are the best materials to use for furnace linings.

Fatigue

Consider a ductile metal having a typical tensile curve of the form shown in Fig. 4.10, from which we infer that the test specimen breaks when subjected to a stress greater than the maximum R_M. With stresses up to R_1, less than R_M, the specimen is once again in a state where the strain is elastic, the elastic region having been extended by work-hardening after the initial plastic deformation (Fig. 4.11). The metal has a stable atomic structure if stresses less than R_1 are applied: the interatomic distances between all atoms undergo only small elastic variations (of the order of 0.1 per cent) and are therefore completely reversible.

This is true if the stress is steady or slowly varying, but a new effect appears when the component is subjected to an alternating load with a regular frequency of oscillation, an effect that causes problems in many technical applications. In rotating machinery, for example, each component experiences a regular cycle of stresses, while vibrations or cyclic loads are very common in mechanical equipment. Experience then shows that a component subjected to an alternating stress of magnitude much less than the static ultimate strength R_M may break after a certain number of cycles in service, often a very large number. This is called **metal fatigue**.

To characterize the effect more precisely, a test specimen is subjected to a cyclic load of known amplitude and frequency, 100 Hz for example. The number of cycles before fracture is measured as a function of the maximum applied stress. The test may continue for tens of millions of cycles and might therefore take a very long time.

With mild steel, it is observed that there is a maximum stress, the **endurance limit**, below which the specimen does not break, however many cycles there are. Its value is approximately half the static tensile strength. As the stress is increased above this limit, the number of cycles before failure decreases, at first rapidly and then more slowly (Fig. 4.42).

For other metals, such as light alloys, the endurance limit is not well-defined. As the stress is reduced, fracture is delayed for longer and longer times but, even for stresses that are very small in comparison with the static limit, the specimen may break after a sufficiently large number of cycles.

It is clear that fatigue is of considerable practical importance. In order to guarantee complete safety when using steel components under cyclic

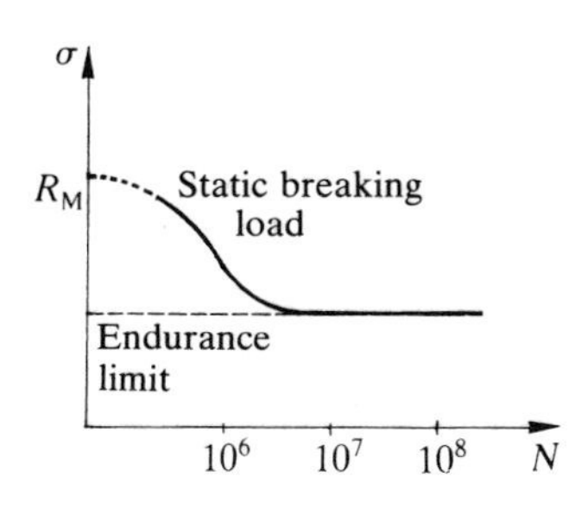

Fig. 4.42. Metal fatigue. The Wöhler diagram or *S–N* curve for mild steel, i.e. the stress plotted against the number of cycles before failure.

conditions, we have to be prepared to reduce the maximum permitted loads to 50 per cent of those allowed under static conditions. With light alloys, the situation is still more serious: even if we impose severe restrictions on the maximum operating loads, predetermined limits must be laid down for the useful life of a machine if all risk of accident is to be avoided.

Atomic mechanism of metal fatigue

What atomic structures are basically responsible for fatigue fracture? As with any other type of fracture, the immediate cause is the presence of microcracks in the metal which grow, slowly at first during the cyclic loading, and then catastrophically when the crack reaches a critical size. What needs to be explained is why microcracks can develop under cyclic stresses, whereas they do not do so when the stresses are applied statically.

When the surface of a metal in the initial stages of fatigue (i.e. after only about 5 per cent of the number of cycles likely to produce fracture) is examined under the microscope, their crystals are seen to be crossed by slip bands, described as 'persistent' since they cannot be removed by light polishing. The bands are thin slices of the material in which a large number of slip lines are concentrated. It follows that dislocations must have been propagated along the planes in these bands, and that this must have involved them in a to-and-fro movement since the stresses were cyclic. The oscillating dislocations must cause the metal in the slip bands to receive a considerable 'shake-up', and it can well be imagined, after our experiment on macroscopic objects, that this might produce a type of atomic motion that would not occur in a metal subjected to a constant or steadily increasing stress. It is by starting from this idea that physicists have attempted to establish models for the atomic motion which are confirmed by direct observation.

If the fatigue test is interrupted and a tensile test is carried out on the fatigued metal, a greater degree of hardening or work-hardening is often observed after a large number of cycles than at the beginning of the test. Fatigue hardening, as it is called, is likely to be caused by the presence of clusters of vacancies hindering the propagation of dislocations: these newly-created micro cavities are initiated by the fatigue process but not by a mere increase in plastic deformation. This interpretation is confirmed by the fact that fatigue hardening is minimal at low temperatures, where vacancies cannot be involved since they are not mobile enough. It is now a well-established and seemingly quite general fact that there are, in the persistent slip bands, displacements of material on a microscopic scale (of the order of a micrometre). An examination of the free surface of materials under the microscope reveals both small voids or crevices, known as

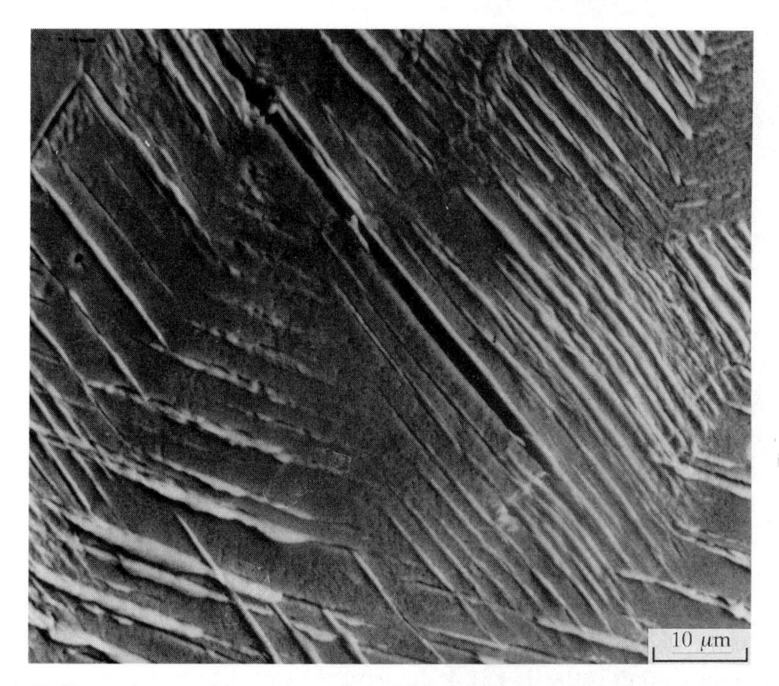

Fig. 4.43. Extrusions and intrusions along slip planes during a fatigue test (scanning electron micrograph, Laboratoire de mécanique, ENSMA, Poitiers).

intrusions, between the slip planes and eruptions or thin tongues of material, known as **extrusions**, growing out of the surface (Fig. 4.43). All such displacements of material can be explained in terms of oscillating slip bands.

As a result, a long sequence of alternating stress cycles, by activating motion on an atomic scale, induces the formation of a large number of microcracks. It is true that there is no direct proof of their existence. It is impossible to obtain electron microscope images of them because of the difficulty of preparing suitable samples. A special method using neutron diffraction (small angle scattering) seems capable of detecting microcavities within the metal but no decisive results have so far been obtained.

All that we can do, therefore, is to observe the ultimate effect of the microcracks: fatigue fracture. Only a few of the initial microcracks develop any further and to be capable of this they must have an appropriate shape and be located in a region where there is a high stress concentration: this often occurs at the surface of the component, e.g. near an irregularity with a sharp profile.

Examination of a component that has suffered fatigue fracture suggests the failure mechanism. Figure 4.44 shows the characteristic appearance of a fatigue fractured surface with two quite distinct regions. Over the first,

Fig. 4.44. Fatigue fracture surface on an engine shaft. The crack initiation appears on the left and the final fracture zone on the right (IRSID).

the metal is rather smooth with successive grooves centred on a point in the surface which is evidently the origin of the fracture. A microcrack would originally have been located at this point, initiated by what is often quite a minor event at the surface and subsequently extended by successive cycles of the alternating stress. The initial crack progresses by joining up with other cracks. In this way, the metal is slowly damaged over a very large number of cycles. The visible contours mark the successive stages at which the opening and closing of the tip of the crack encountered local resistance of varying magnitude during its propagation. Then, when the crack had become large enough, it initiated the sudden failure of the whole specimen. This is the second very rapid stage, corresponding to the second distinct region of the fracture surface: it resembles normal brittle failure, with facets having orientations that vary from grain to grain according to the directions of the cleavage planes in the different crystals.

How can fatigue strength be improved?

It is quite clear that fatigue depends on many details of the structure of a component which we cannot determine: two specimens produced in as identical a manner as possible are, on a microscopic scale, as different as two individuals from the same species of living beings. It follows that we can only make statistical predictions of the behaviour of a single component. Under such conditions and for such metals, there is a risk that the 'disease' of fatigue will be fatal: under other circumstances the possibility of an accident can reasonably be excluded.

Fatigue fracture generally starts at the surface of the component, so that the condition of the surface is an important factor. Systematic removal of any irregularities that can create stress concentrations increases the endurance limit of the metal: this is the effect produced by deep polishing or surface hardening (by cementation, nitriding, or shot peening). In designing the outline of a component, sharp notches or edges that are not carefully rounded must be avoided. A dramatic demonstration of the need for such precautions was provided by the accidents with Comet aircraft in 1950. These, the first commercial jets, had passed all the traditional tests and the first regular flights had been accident-free. Nevertheless, two aircraft broke up in flight after three thousand hours of service. A long enquiry, which was a model of its kind, showed that the cause of the accidents was the failure of the cabin and that the source of the fatal crack was the edge of a cabin window with a defective design. The cruel conclusion is that the accidents could easily have been avoided.

Apart from taking care in the design and surface treatment of components subjected to cyclic loading in service, there is little we can do to avoid metal fatigue. In many cases, it is the main effect restricting the technological potential of a metal. The endurance limits of all commercial products are established by carefully controlled systematic tests and in this way it has been possible to discover the best compositions and the most effective heat treatments for alloys. However, there is no known theoretical recipe for reducing the susceptibility of a material to metal fatigue.

Furthermore, there is at present no reliable method of assessing the state of a component in service as far as fatigue is concerned. Of course, a minute examination of the surface of a component towards the end of its useful life might sometimes reveal the crack that will soon cause failure. But there is no non-destructive test for distinguishing a component which can still withstand millions of cycles without damage from one that might well fail after only a few thousand more cycles.

Some comments on the mechanical properties of metals

We have made a point of providing a general overview of the mechanical properties of metals. We have emphasized only a few important characteristics and have deliberately omitted a detailed description of the properties of the various metallurgical products in current use.

A catalogue of that sort would demonstrate the wide range of properties possessed by alloys, yet the number of different metals incorporated in them is very restricted. From the complete list of metals, we have to exclude those that are too scarce in the Earth's crust (and therefore too expensive) and those that are affected too greatly by the atmosphere. In fact, a mere three metals, iron, aluminium, and copper, form the basis of almost all the alloys used in mechanical construction.

If it were true that, overall, the properties of an alloy are a linear combination of those of its pure constituents, the range of possibilities in metallurgy would be extremely restricted. That is certainly what happens in the case of properties like density that are independent of atomic structure. Whatever the structure of an alloy, the volume occupied by all the atoms is directly related to the sum of the volumes occupied by the different types of atom. Consequently, adding 1 per cent of a new element only brings a contribution from its own density of about 1 per cent to the overall density. All steels have relative densities of 7.9 to within 5 per cent.

On the other hand, the failure load of titanium or zirconium is profoundly affected by the presence of a few parts per million of oxygen. Or again, the elastic limit of a given steel may be doubled by a minimal change in heat treatment.

In this chapter, we have revealed the underlying cause of such behaviour in mechanical properties. The deformation or failure of a metal is not determined only by the composition and the crystal structures of the various phases that are present, but depends on the existence of defects in the solid (dislocations, microcracks, etc.) and their reaction to external stresses. The entities involved are details in the real structure of the metal: local deformation of the crystal lattice, the texture of grains and grain boundaries, and so on. It becomes clear that the properties depend on the treatment to which the metal is subjected during its production and that trace elements can have an effect out of all proportion to their very low concentration, rather like a catalyst.

It has proved possible, thanks to alloys, to create materials with varied properties that are much superior to those of the pure constituents. The advances made in mechanical construction through the use of steels and light alloys would not have been possible with pure iron and pure aluminium on their own. Another great advantage of metallic alloys is the ease with which they can be prepared by mixing the melted components and then cooling the liquid to give a homogeneous mass.

Lengthy and thorough laboratory research has produced better products. But the greater the progress, the more the conditions for the preparation of a product become 'specialized'. Thus, minute changes in industrial production can seriously alter the quality of a metal: for example, 'poisoning' by a minor impurity coming from a new source of raw material or a slight drift in the conditions of treatment. An aluminium factory was manufacturing zip-fasteners without any trouble, when suddenly it was noticed that cracks were occurring where the hooks gripped the cloth. The number of rejected items became prohibitive and it needed long and difficult research to find the nature of the fault responsible and the possible remedies.

Maintaining the quality of an alloy is an operation requiring very high standard of control at every stage of production. Monitoring must be thorough and accurate and, to be effective, it must be based on reliable fundamental knowledge. As a result, theoretical solid state physicists are not only involved in laboratory research directed towards the discovery of new materials but are also concerned with the industrial quality control aimed at ensuring that the standard of the finest products is maintained.

5

Diffusion

The structural elements (atoms, ions, or molecules) in all matter, whatever its state, are subject to thermal agitation. The thermal energy per particle associated with this agitation at a temperature T is of the order of $10^{-4} T$ eV.

In gases, where the molecules are widely separated and the interactions between them very small, the thermal energy is purely kinetic: molecules have mean velocities due to thermal agitation of the order of 100 metres per second. The velocities have random directions and are constantly changing as a result of collisions.

In a crystalline solid, each atom or ion is associated with a well-defined site in the crystal lattice: thermal motion then consists of oscillations about the theoretical site, which becomes the mean position of the vibrating atom.

We thus have two very different situations. *In gases*, because the translational velocities of the molecules are high and in random directions, there is continual mixing of material. This is **diffusion**. A given molecule gets further and further away from its initial position but, because the motion is completely random, its drift velocity is much less than its instantaneous velocity and is only of the order of a few millimetres per second at room temperature. Nevertheless, after a sufficiently long time, a molecule will have occupied every point in the vessel containing the gas. If two vessels containing different gases are put into communication with each other, the gases will mix and become homogeneous within a few minutes.

Such spontaneous diffusion also occurs in a *liquid* (a condensed and disordered state of matter), but molecular motion is impeded by the interactions between the molecules which are quite closely packed together. Diffusion rates in liquids are low compared with those of gases (in a ratio of 1 to 1000). Experiment shows that two miscible liquids, initially separate, interdiffuse and become homogeneous only very slowly. A mixture is made in practice by shaking the containing vessel, i.e. creating convection.

Diffusion would be impossible in a *perfect crystalline solid*, since the atoms remain imprisoned by their neighbours in fixed positions in spite of their thermal agitation. However, experiment shows that, in *real solids*, diffusion does occur, although it is very slow in comparison with that observed in liquids. Thus, if two blocks of metal, one of lead and the other of gold, are pressed strongly together after the surfaces in contact have been scrupulously cleaned and smoothed, lead atoms are found to have penetrated into the gold and vice versa. After being held at a temperature of 300 °C for several days, however, the atoms have travelled a distance of less than a micrometre.

We are led to the idea that diffusion can take place in real crystals because of the existence of defects, playing a vital role here as they do in other properties of solids. The defects responsible for diffusion are **vacancies** (SM, p. 107): in spite of their rarity (there is never more than one atom in a thousand missing), they have an observable macroscopic effect.

Diffusion is of considerable importance in the physics of solids, especially in metallurgy. It determines what physical transformations and chemical reactions can occur in the solid state and how they proceed. It is found that, at easily attainable temperatures, diffusion enables structural changes to take place in alloys over times that are quite short on the human time scale. It is obvious that industrial heat treatment of alloys at 500 °C would not be economically feasible if it had to last a month, and that it would not even be considered if it needed to last a year. Times of the order of a few minutes or even a few hours, on the other hand, are acceptable. The preparation of components in the semiconductor industry also involves diffusion: here again, it is only because they can be produced in 'short' times that they can be marketed at an economic price.

Experimental data on diffusion

The simplest phenomenon is **self-diffusion**, i.e. the motion of atoms inside a pure solid. But how can such motion be revealed if all the atoms are identical? We achieve this in fact by 'marking' a small proportion of them: the normal atom is replaced by a radioactive isotope which has exactly the same physical and chemical properties, but which can be tracked by detecting the radiation it emits when it decays. The motion of a collection of atoms can be followed on this way.

The type of diffusion that is important from a practical point of view is **interdiffusion** in solids containing several different types of atom. We shall restrict ourselves to the case of a solid solution with two components. The A and B atoms are at the nodes of a single lattice whose sites may be occupied randomly by either type of atom. This is a situation that occurs in metals: the solubility of the components is often only partial, but below the solubility limits the effects remain simple. The basic process is an interchange of A and B atoms between the lattice sites they occupy. We shall not consider the more complex cases where there are intermediate compounds A_mB_n with structures that are different from that of the pure metals.

No diffusion will occur if the elements are mutually insoluble: this is seldom the case with metals, but is common in minerals and other inorganic crystals.

The diffusion conditions depend on the geometrical arrangement of the metals present. Thus, when a grain of B is included in A, the atoms of B will spread out through A around B. The diffusion will be described by the variation with time of the concentration of B (or A) atoms at each point in the whole system. In order to extract data relevant to diffusion itself and eliminate geometrical complications, it is worthwhile choosing as simple an arrangement as possible. A bar of metal A is pressed firmly against a bar of B so that the contact between them is perfect (Fig. 5.1). The observable diffusion is then in one dimension, along the common axis of the bars. The concentration of B is a function of the distance from the initial interface and of the time from the beginning of the diffusion.

Before diffusion begins, the number of atoms of B per unit volume increases abruptly from 0 to c_0 at the boundary between the metals A and B. The metal B then penetrates A and vice versa, and this causes the sharp step in the concentration curve to become rounded off and the slope at the origin to become less and less steep. If there was total mutual solubility of A and B, the equilibrium state would be a uniform distribution of the A and B atoms. In practice, the situation always remains a long way from this theoretical equilibrium state.

Two main features are revealed by the experiment:

1. Diffusion in solids is a slow process, so that its progress can easily be observed. To specify what is happening more precisely, we consider the transition zone containing both A and B atoms with relative concentrations lying between 25 per cent and 75 per cent. Table 5.1 relates to a typical case of intermetallic diffusion: nickel into iron at 1200°C. It gives the thickness of the transition zone for various diffusion times. After several days, the transition zone is about a tenth of a millimetre thick.

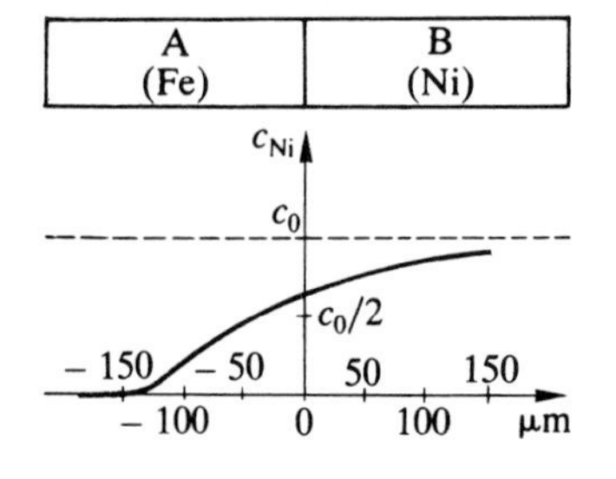

Fig. 5.1. Diffusion of nickel in iron held at 1200°C for 48 hours. c_{Ni} is the number of nickel atoms per unit volume.

Table 5.1. Diffusion of nickel into iron at 1200 °C

Diffusion time	Thickness of transition zone
45 minutes	15 μm
3 hours	31 μm
12 hours	63 μm
2 days	126 μm
8 days	250 μm

2. The diffusion rate increases very rapidly with temperature. Although almost zero at room temperature, it becomes appreciable near the melting point of the metals concerned. To give some idea of the magnitude of the effect, we can say that the diffusion rate approximately doubles for every 10° rise in temperature.

The diffusion coefficient

Consider a block of a solid solution of two metals A and B, which are soluble in all proportions. If the A/B solid solution is initially heterogeneous, the system will tend spontaneously to become uniform as is predicted by thermodynamics. Diffusion is the mechanism by which this homogenization takes place: a flux of B atoms leaves the regions rich in B and makes for regions poor in B: it is the same with A. The flux is greater when the variation in the concentration of B atoms with distance is steeper, i.e. when the concentration gradient is higher.[†]

The transport of matter is analogous to the transport of heat in a body at a non-uniform temperature or the transport of electric charge in a conductor whose potential is not uniform. In the latter case, we know that the flux of electric charge is proportional to the potential gradient or electric field. In the other cases, the flux is proportional to the gradient of the quantity playing the same role as the electric potential: temperature in the case of heat and concentration in the case of diffusion.

To go beyond this qualitative approach, we shall define the parameter which gives a measure of the diffusion that takes place: the **diffusion coefficient**. We return to the one-dimensional diffusion in a bar, where the

[†]Since the concentration c at a point is a function of three coordinates x, y, and z, the gradient of c is a vector with components $\partial c/\partial x$, $\partial c/\partial y$, $\partial c/\partial z$. In a one-dimensional system, c is a function of x only, and the gradient is a scalar, the derivative dc/dx.

concentration c giving the number of B atoms per unit volume depends only on the x coordinate. The diffusion is governed by **Fick's law**: the number of B atoms crossing unit area per unit time in the x direction at a cross-section of the bar located at x is proportional to the gradient of the concentration, i.e. to its derivative with respect to x:

$$J(x) = -D\,\mathrm{d}c/\mathrm{d}x.$$

The negative sign means that diffusion tends to make the concentration more uniform, i.e. to reduce $\mathrm{d}c/\mathrm{d}x$.

D is the diffusion coefficient. It plays a similar role to the electrical conductivity or the thermal conductivity in the transport of electricity and heat respectively.

The diffusion coefficient, usually measured in $\mathrm{cm^2 s^{-1}}$, is generally negligible at room temperature (with a value of about 10^{-30}) and is of the order of 10^{-8} at the melting point of metals.

Interpretation of the diffusion coefficient on an atomic scale

Diffusion of an atom in a pure crystal

The fundamental feature is the fact that an atom has a chance of changing places with any one of its immediate neighbours: the mobility of an atom will be measured by the average number Γ of jumps that it makes per second. Suppose an atom is radioactively marked in a pure element. We consider the case of a crystal with a simple cubic lattice. Although this is not a realistic example, since no element crystallizes with this type of lattice, by using it we avoid geometrical complications in the arguments and we end up with a result that can be generalized to the lattices of real metals (face-centred cubic, body-centred cubic, etc).

After n jumps, and thus after a mean time $t = n/\Gamma$, the marked atom is at a distance from the origin given by a vector $\boldsymbol{L}(t)$. Because the diffusion is completely isotropic, any displacement has an equally probable displacement in the opposite direction, so that the mean value $<\boldsymbol{L}(t)>$ is zero. On the other hand, the magnitude of the mean distance travelled is not zero. The atom follows a zigzag path, consisting of n steps each of length a. It therefore travels a distance na but the root mean square displacement of the atom is $a\sqrt{n}$.[†] Thus:

$$<L(t)>^2 = na^2 = \Gamma a^2 t.$$

[†]We are using the solution to a classic problem in the probability calculus, the random walk (constant steps in random directions).

It is therefore the square of the distance covered by the atom on average that is proportional to the time of diffusion.

When the A and B atoms form an ideal solid solution, i.e. when A and B play the same role, an isolated B atom in the A crystal diffuses like the marked A atom and it is the same for an isolated A atom in a B crystal.

Fick's law

We have just evaluated the displacement of an atom due to its random motion in a solid. We must now show what consequences this has in one-dimensional diffusion in a cylindrical bar in which the concentration of B is a function of the x coordinate, and thus obtain Fick's law.

We consider a highly simplified situation: a single crystal of a solid solution of A and B with a simple cubic lattice, so oriented that one of the cubic axes is directed along the axis of the bar (Fig. 5.2). The concentration is uniform over each atomic plane but varies from one plane to another. Although the atoms move by random jumps in three dimensions, the observable result is diffusion in one dimension: the variation with time of the concentration of B atoms in successive lattice planes.

Let two atomic planes P_1 and P_2 have coordinates x and $x+a$. A B atom will jump from P_1 to P_2 with a frequency $\Gamma/6$: this is because it is one of the six elementary jumps possible for this atom. The number of B atoms per unit area of P_1 is ca (because the average number of B atoms per unit cell of volume a^3 is ca^3 and there are $1/a^2$ unit cells per unit area). The number of atoms passing from P_1 to P_2 per second is therefore $c(x)a\Gamma/6$. In the same way, the number of B atoms jumping from P_2 to P_1 is $c(x+a)a\Gamma/6$. The two contrary currents do not cancel each other since there is a different concentration in the two adjacent planes.

The flux of B atoms will be:

$$J(x) = a\,\frac{\Gamma}{6}\,[c(x) - c(x+a)] = -\,\frac{1}{6}\,a^2\Gamma c'(x).$$

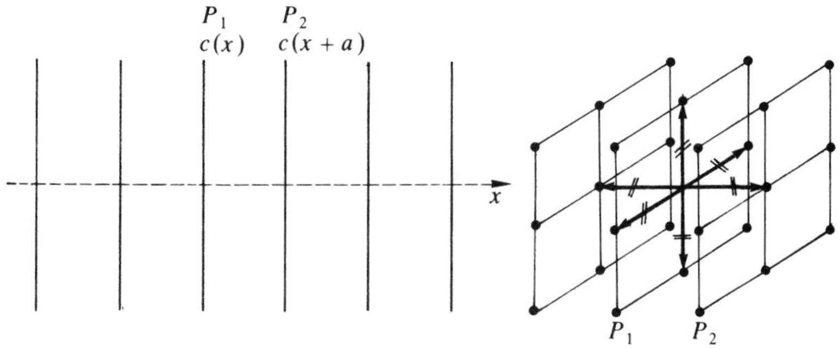

Fig. 5.2. Calculation of the diffusion coefficient.

By identifying this result with Fick's law, we obtain an expression for the diffusion coefficient in terms of a characteristic of the atomic model, the frequency with which an atom jumps into a neighbouring site:

$$D = a^2 \Gamma/6$$

or

$$D = a^2 \Gamma_S$$

where the symbol $\Gamma_S = \Gamma/6$ denotes the jump frequency along the direction of the diffusion.

Diffusion is the result of completely disordered atomic motion. In the case of self-diffusion in a pure solid or of interdiffusion in an ideal solid solution, there are no 'forces' pushing the atoms towards uniformity of concentration: it is a spontaneous process, in conformity with thermodynamics (the increase of entropy in a system with constant internal energy).

Variation of the diffusion coefficient with temperature

An atom jumps from one site to a neighbouring one because it is vibrating about its mean position as a result of thermal agitation. In general (and that means almost always), an atom that is displaced from its site is taken back to it by the action of the neighbours forming a cage around it. However, the amplitude of vibration is not constant, but fluctuates randomly. A fluctuation may occur that is so great that the atom acquires sufficient energy to enable it to force its way through the barrier that keeps it in place: it will then be found in the neighbouring site.

Consider a mechanical analogy of this: a ball is rolling on an undulating track with two neighbouring minima M_1 and M_2 (Fig. 5.3). If the movement has only a small amplitude, the ball will oscillate in the valley M_1. If, however, it is activated by an energy greater than mgh, where h is the height of the hill between M_1 and M_2, it will reach the hill, pass over it and fall into the valley M_2. In a similar way, an atom vibrating around a site, will pass into the neighbouring site if it is activated by an energy of 'liberation' Q_a. According to Boltzmann's law, the probability that an

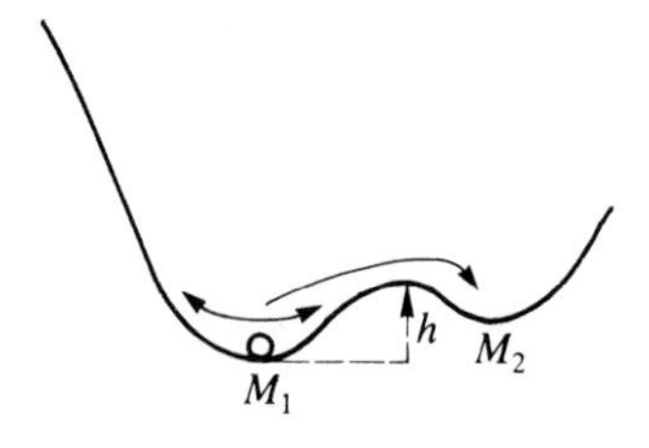

Fig. 5.3. Mechanical analogy for diffusion: a ball on a slide. The activation energy for it to jump from M_1 to M_2 is mgh.

atom has an energy Q_a is $\exp(-Q_a/k_BT)$. Out of the ν vibrations made by the atom per second, the number of jumps Γ will be:

$$\Gamma = \nu \exp(-Q_a/k_BT).$$

Hence:

$$D = \frac{\nu a^2}{6} \exp\left(\frac{Q_a}{k_BT}\right) = D_0 \exp\left(-\frac{Q_a}{k_BT}\right) = D_0 \exp\left(-\frac{Q}{RT}\right).$$

Q_a is the energy per atom and Q the energy per mole (the ratio $Q_a/k_B = Q/R$ has the dimension of a temperature).

This is Arrhenius's law (Fig. 5.4) describing the variation of D with temperature, as it does for many chemical reactions that are initiated when a potential barrier is surmounted. The exponential in the expression for D accounts for the very rapid variation of the coefficient with the temperature.

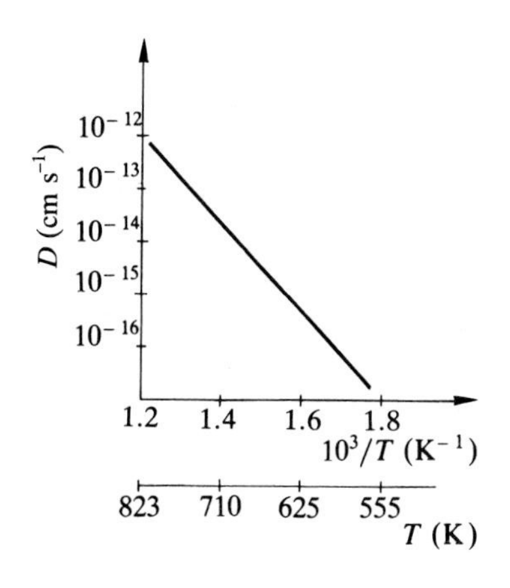

Fig. 5.4. Coefficient of self-diffusion for gold as a function of temperature. Confirmation of Arrhenius's law with $Q/R =$ 20200 K.

Experimental measurement of the diffusion coefficient

How can we predict the curve showing the distribution of concentration in a diffusion experiment from the value of the coefficient D? Conversely, how can we deduce the value of D from an experimentally determined concentration curve?

Once again, we use the case of one-dimensional diffusion along a bar formed from two metals A and B welded together. The concentration c of B atoms is a function of two variables: the distance from the interface and the time of diffusion. The mathematical tool we use is a partial differential equation, which turns out to be the transposition of the equation for heat flow (Fourier's equation) to the case of diffusion. Integration of this equation gives $c(x, t)$ as a function of the diffusion coefficient D. Conversely, if $c(x, t)$ is measured, D can be calculated.

Box 21

The diffusion equation: solution to the bar problem

Consider the problem of one-dimensional diffusion along a bar of constant cross-section S. The concentration of B atoms $c(x, t)$ depends both on the space variable x and the time variable t. We assume that it is independent of any other space variables, i.e. that it is constant over any plane cross-section of the bar. We have seen that the flux of atoms diffusing across unit area at the coordinate x is given by Fick's law:

$$J(x) = -D\partial c(x, t)/\partial x.$$

(Remember that the notation $\partial c/\partial x$ indicates the partial derivative of c with respect to x, i.e. the derivative with respect to the single variable x, the time being constant.)

We shall now complement this equation by another which connects J and c. This is the equation of **conservation** or of **continuity**, which simply expresses the balance in the quantities of matter involved. Consider a slice of the bar between the coordinates x and $x + \Delta x$ (Fig. 5.5). During an interval of time Δt, $J(x)S\Delta t$ atoms arrive at the left-hand surface of the slice and $J(x + \Delta x)S\Delta t$ leave the right-hand surface. If $J(x)$ and $J(x + \Delta x)$ are different, the total number of atoms contained in the slice Δx will vary: there is an accumulation if $J(x) > J(x + \Delta x)$, i.e. more will arrive than leave. The excess number of atoms is:

$$[J(x) - J(x + \Delta x)]S\Delta t = -\frac{\partial J(x)}{\partial x} S\Delta x\Delta t.$$

Dividing by the volume of the slice, $S\Delta x$, we obtain the variation in concentration during the time Δt. Hence:

$$\frac{\partial c}{\partial t} = -\frac{\partial J}{\partial x}.$$

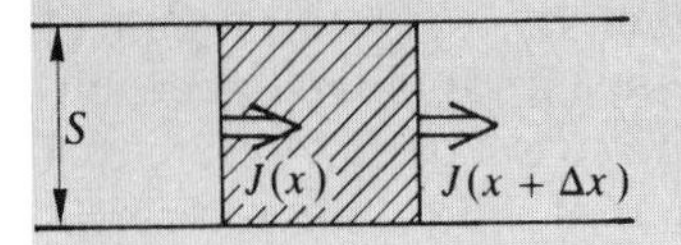

Fig. 5.5. Diffusion along a selection of a bar in the establishment of the diffusion equation.

Combining this equation with Fick's law, we obtain:

$$\frac{\partial c}{\partial t} - \frac{\partial}{\partial x}\left(D\frac{\partial c}{\partial x}\right) = 0.$$

By taking the case in which D is independent of x (i.e. of c), the following equation, known as the **diffusion equation**, is obtained:

$$\frac{\partial c}{\partial t} - D\frac{\partial^2 c}{\partial x^2} = 0.$$

This is an equation relating the partial derivatives of $c(x, t)$ and which enables $c(x, t)$ to be found by integration if the initial 'profile' $c(x, 0)$ is known.

We give the solution of the equation for a simple case in which:

$$\begin{cases} c(x, 0) = c_0 \text{ for } x < 0 \\ c(x, 0) = 0 \text{ for } x > 0. \end{cases}$$

We introduce the dimensionless variable $u = x/2\sqrt{Dt}$ and see whether c might be a function of this variable only. To do that, we first calculate the partial derivatives of c with respect to x and t in terms of the partial derivatives of c with respect to u:

$$\frac{\partial c}{\partial x} = \frac{\delta u}{\delta x}\frac{dc}{du} = \frac{1}{2\sqrt{Dt}}\frac{dc}{du}.$$

Differentiating again:

$$\frac{\partial^2 c}{\partial x^2} = \frac{1}{4Dt}\frac{d^2 c}{du^2}.$$

Similarly:

$$\frac{\partial c}{\partial t} = \frac{\partial u}{\partial t}\frac{dc}{du} = -\frac{x}{4\sqrt{D}t^{\frac{3}{2}}}\frac{dc}{du}.$$

Substituting these expressions in the diffusion equation, we have:

$$-\frac{x}{4\sqrt{D}t^{\frac{3}{2}}}\frac{dc}{du} = \frac{1}{4t}\frac{d^2 c}{du^2}.$$

It can be seen that the derivative $c' = dc/du$ satisfies the differential equation:

$$dc'/c' = -2u du$$

which can be integrated to give:

$$c'(u) = a\mathrm{e}^{-u^2}.$$

This can be integrated once more to give:

$$c(u) = a\int_0^u \mathrm{e}^{-v^2}\mathrm{d}v + b.$$

The initial conditions rewritten with the variable u are:

$$\lim_{+\infty} c = 0 \quad \text{and} \quad \lim_{-\infty} c = c_0.$$

Using the fact that $\int_0^{+\infty} \mathrm{e}^{-v^2}\mathrm{d}v = \frac{\sqrt{\pi}}{2}$, we obtain:

$$b = c_0/2 \text{ and } a = c_0/\sqrt{\pi}.$$

Thus:

$$c(u) = c_0\left(\frac{1}{2} - \frac{1}{\sqrt{\pi}}\int_0^u \mathrm{e}^{-v^2}\mathrm{d}v\right).$$

The term:

$$\mathrm{erf}\,(u) = \frac{2}{\sqrt{\pi}}\int_0^u \mathrm{e}^{-v^2}\mathrm{d}v$$

is usually called the *error function* and is an integral of the Gaussian distribution function. Returning to the variables x and t, we thus have:

$$c(x,t) = \frac{c_0}{2}\left[1 - \mathrm{erf}\left(\frac{x}{2\sqrt{Dt}}\right)\right].$$

Figure 5.6 gives the curve of $c(x, t)$ as a function of x at different times and for a diffusion coefficient equal to $0.16 \times 10^{-8}\,\mathrm{cm^2 s^{-1}}$.

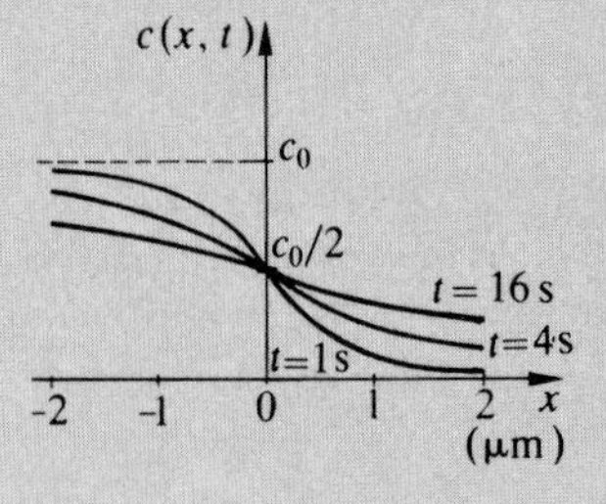

Fig. 5.6. Solution of the diffusion equation for a bar: $c(x, t)$ for $t =$ 1s, 4s, and 16s and with $D = 0.16 \times 10^{-8}\,\mathrm{cm^2 s^{-1}}$.

The main experimental difficulty is the measurement of the local concentration with sufficient geometrical accuracy. The transition zone after diffusion is always very thin, of the order of a tenth of a millimetre. Measurements will therefore only be meaningful if the analysis is made on slices which are about a micrometre thick. The Castaing electron microprobe makes this possible without affecting the specimen (Fig. 5.7). There are many other techniques, but they involve the destruction of the specimen to be analysed.

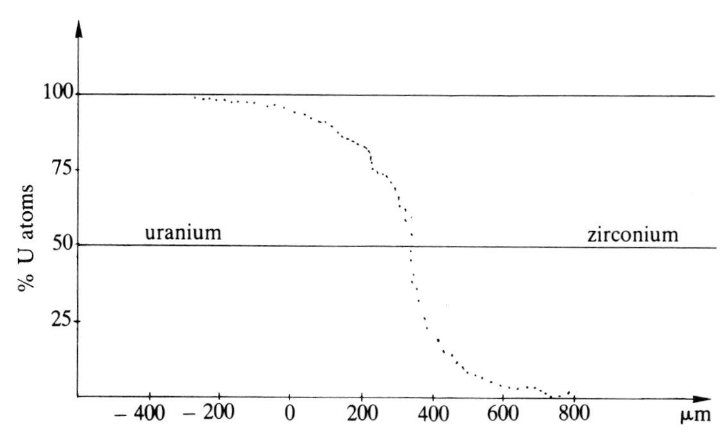

Fig. 5.7. Measurement of the diffusion profile with the Castaing electron microprobe: a U–Zr pair held at 1000°C for 48 hours (by kind permission of J. Philibert).

Experimental measurements are in good agreement with the main theoretical results we have just given. Firstly, we often find that the curve showing the measured concentration in the interdiffusion of a pair A–B is described by the equation derived from Fick's law, i.e. that a value for D can be found for which the theoretical concentration curve agrees well with experiment. In particular, it is confirmed that the thickness of the transition zone (Table 5.1) is proportional to the square root of the time of diffusion. If D is determined at various temperatures, the diffusion coefficient is found to obey Arrhenius's law: the logarithm of D varies linearly with $1/T$ (Fig. 5.4) and the slope of the line gives the parameter $Q_a/k_B = Q/R$.

For metals, this parameter lies in the range from 10000 to 30000 K. An empirical relationship in common use that gives more precise values is:

$$Q/\mathrm{R} \approx 17T_f$$

where T_f is the melting point. There is one important exception: silicon has an activation energy which is about twice that of a normal metal.

Finally, in the theoretical formula:

$$D = \frac{\nu a^2}{6} \exp\left(-\frac{Q}{RT} \right).$$

the measured values of D give a value for ν which is very much the same order of magnitude as the frequencies of atomic vibrations (10^{12} to 10^{13} Hz).

For a metal in its solid state at the melting point, the diffusion coefficient is about $10^{-8} cm^2 s^{-1}$. For the metal in liquid form at the same temperature, the coefficient is about 10^{-5}. This indicates how much easier it is for diffusion to take place in matter in a disordered state. Silicon at its melting point has a coefficient of only 10^{-12} because of its unusual value for the activation energy of atomic jumps. This property makes the operation of doping silicon (or germanium for that matter) by diffusion a very long process. On the other hand, it ensures that components with zones having different doping levels are stable, even at operating temperatures above room temperature.

The role of vacancies in diffusion

So far, we have assumed that atomic jumps are the cause of the interchange of atoms between lattice sites without specifying the mechanism by which such an exchange might occur. We have simply characterized it by its activation energy, which can be measured experimentally from the variation of D with temperature. We must now attempt to find out whether the activation energy predicted for the atomic jump by an atomic model agrees with the value obtained experimentally.

In a perfect crystal, where the atoms are as close-packed as possible, the jump of an atom into a neighbouring site must be accompanied by the jump of the dislodged atom in the reverse direction: there is no room for an interstitial atom in a close-packed lattice. When two neighbouring atoms exchange sites, they must pass through an intermediate position in which the lattice is greatly distorted around them (Fig. 5.8). The energy required to reach this situation has been calculated and is of the order of 5 eV per

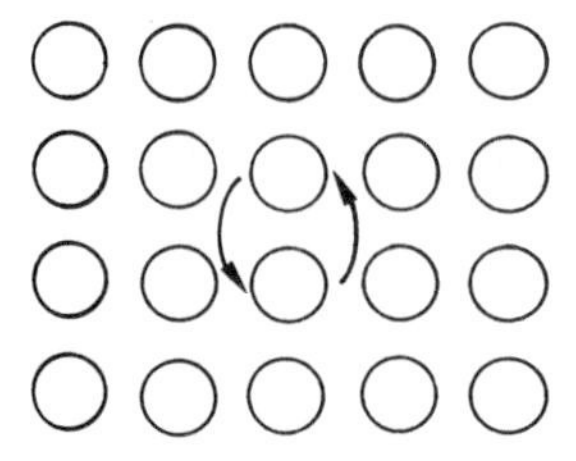

Fig. 5.8. Direct interchange of neighbouring atoms.

atom, or more than twice the experimental activation energy (about 2 eV). This would correspond to a value for D that is 10^8 times smaller at 1200 K than the measured coefficient. In other words, to reach a given state of diffusion at a given temperature would require a time that was 10^8 times longer than is actually observed. The process is therefore impossible. Other mechanisms for the interchange of atoms in a perfect lattice have been suggested, with equal lack of success.

The difficulty is resolved by invoking **vacancies** (SM, p. 107), which occur in all crystals at high temperatures. An atom adjacent to a vacancy can become lodged in the empty site by leaving its own site empty (the vacancy is then said to migrate). Because there is plenty of space in a vacancy, this movement requires a much smaller activation energy than that needed for a displacement in a close-packed crystal. It is easy to see from simple models (e.g. SM, pp. 108–109) how vacancies and their migration make it possible for atoms to move over large distances.

The mechanism of **vacancy-assisted diffusion** can be subjected to an accurate calculation. First of all, the number of vacancies at equilibrium at a given temperature is known: the probability that a site will be vacant is proportional to $\exp(-E_f/k_B T)$, where E_f is the energy of formation of a vacancy. We also know that the probability of a vacancy migrating to an adjacent site is of the form $\exp(-E_d/k_B T)$, where E_d is the activation energy for the displacement of the vacancy. The two quantities E_f and E_d can be measured using experiments that are independent of the phenomenon of diffusion.

There are two conditions to be fulfilled if an atom is to jump: (1) there must be a vacancy at a neighbouring site and (2) the vacancy must migrate. The probability that such a jump will occur is therefore the product of the two probabilities, i.e.

$$\exp\left(-\frac{E_1}{k_B T}\right)\exp\left(-\frac{E_d}{k_B T}\right) = \exp\left(-\frac{E_f + E_d}{k_B T}\right).$$

In the expression for D (p. 249), therefore, the parameter Q_a will be replaced by $(E_f + E_d)$. The measured values for Q_a and for E_f and E_d provide satisfactory confirmation of the predicted relationship, which is a convincing demonstration of the reality of vacancy-assisted diffusion.

An accurate calculation of the diffusion coefficient based on this mechanism is complicated by the following effect. The marked atom, located at M, jumps into the adjacent vacant site N: there is a chance that the next jump would take the atom from N back to the site M that has become vacant. Both jumps are then 'lost' since we are back in the initial situation. The diffusion coefficient is reduced by such a 'correlation' between two successive jumps.

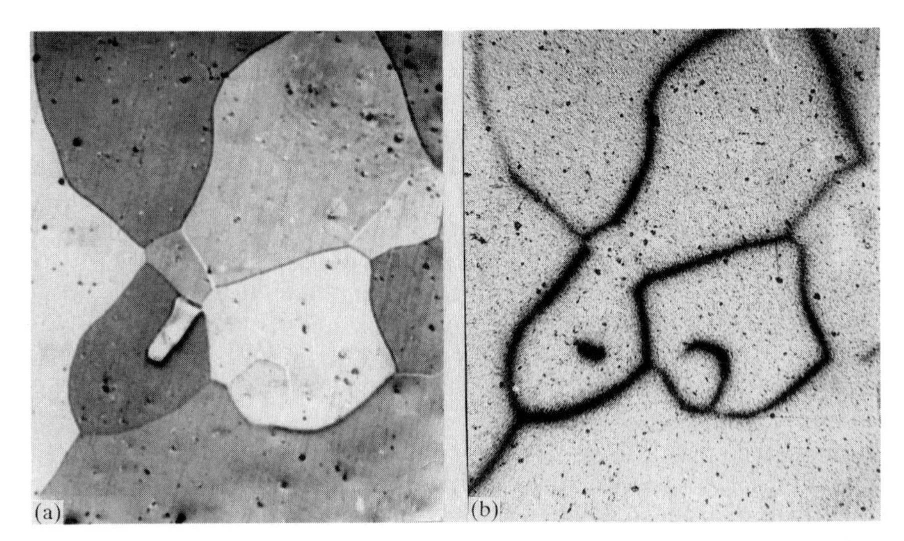

Fig. 5.9. Intergranular diffusion of radioactive Fe in iron: (a) metallographic image; (b) autoradiography. There is no diffusion along the twin boundaries since no disorder occurs there (by kind permission of P. Lacombe).

Any increase in the number of vacancies above their concentration at thermal equilibrium or, more generally, any local distortion reducing the closeness of the packing of the atoms, will accelerate diffusion. In a polycrystalline solid, the grain boundaries are disordered regions: along the common surface between the lattices of adjacent grains, interstices occur between atoms located in irregular positions with respect to both lattices. **Intergranular diffusion** is easier than that taking place through the body of the grain, as is shown by the following experiment. A radioactively marked metal layer is deposited on the surface of the specimen, which is then heated to a temperature that is not high enough for normal diffusion to be appreciable. Intergranular diffusion, however, is already active: marked atoms that have travelled along the grain boundaries can be detected by autoradiography (Fig. 5.9).

Diffusion in real systems

We have envisaged the interdiffusion of one element into another in a very simple case, the ideal solid solution, i.e. such that the replacement of one element by another does not cause any appreciable change in the overall structure or the energy. This was a first approximation, rather unrealistic but nevertheless useful.

To go any further, we must take into account the fact that the atoms are different. Although they may mix in a solid solution, the various sites that an atom can occupy are not equivalent from the energy point of view, some being more favourable than others. It follows that the motion of this atom in the solid is no longer purely random. The atom is subjected by its environment to forces that affect its motion. The result of this is that several of the statements made in this chapter are not always true. Thus the diffusion coefficient of A into B may be different from that of B into A or the coefficient may depend on the concentration, which would prevent Fick's law from being integrated as it was in Box 21.

We shall not tackle these complications, even though they are essential for a correct description of diffusion in real solids, since they require the introduction of *ad hoc* parameters that cannot be derived from atomic models on their own. Once again, we meet the problem raised by insufficient knowledge of interatomic interactions. We should not forget, however, that the simplified model is correct for self-diffusion and that it has enabled us to establish the role of vacancies in diffusion.

We give a few examples of some experimental observations that only a more refined theory of diffusion can explain.

The Kirkendall effect

We return to the experiment involving diffusion between two metal bars in intimate contact, but in a case where the diffusion rates are very different in the two directions. Thus, for a silver–gold pair, silver diffuses more easily into the gold than vice versa: the ratio of the two coefficients at 915 °C is about 3.

The two bars are welded, and at the same time the interface is marked with very fine filaments of an inert metal, tungsten. After treatment at 915 °C for 24 hours, the gold–silver boundary is observed to have moved by 100 μm from the plane marked by the tungsten filaments. The region in the gold that has absorbed silver has a greater volume than that in the silver containing gold: this is the **Kirkendall effect**.

In addition, in a polished cross-section of the bar, small cavities several micrometres across located in the transition zone can be seen under the microscope (Fig. 5.10). These cavities are due to a condensation of vacancies that have become supersaturated following their migration during the diffusion process. The Kirkendall effect is a direct indication that vacancies play a role in diffusion.

Inverse diffusion and the formation of GP zones

In general, diffusion is the atomic mechanism by which a solid can evolve towards its equilibrium structure. When, for a pair of metals, this equilibrium state is a single solid solution phase, the effect of the diffusion is to make the local composition uniform. There are, however, other possibilities.

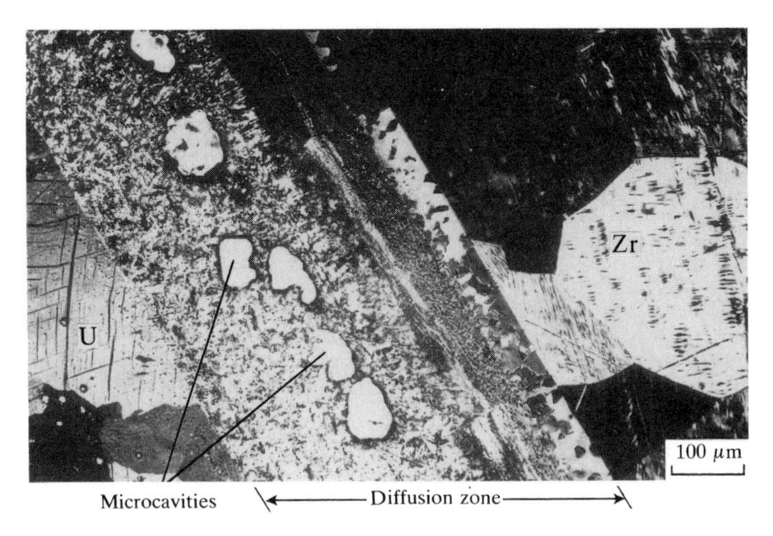

Fig. 5.10 The Kirkendall effect: interdiffusion in the U–Zr pair held at 750°C for 160 hours (by kind permission of J. Philibert).

Consider a **supersaturated** solution A–B obtained as follows: starting from a homogeneous solid solution at a high temperature where the solubility of B is high, we take it suddenly to a low temperature where the solubility of B is much lower. Immediately after the quench, the solid solution is supersaturated since its composition cannot change instantaneously: the equilibrium state is a two-phase alloy of 'precipitates' rich in B included in an A–B solution at the saturation limit and therefore poor in B.

This equilibrium state is reached by means of diffusion. In this case, the diffusion creates heterogeneity: this is **inverse diffusion**, whose mechanism we now examine.

Because of random atomic motion, it is possible for a group of B atoms to gather at a particular point and these, together with a few A atoms, assume the structure of the precipitated phase over a volume amounting to a few unit cells. Since it is the equilibrium structure, the B atoms arriving at the surface of this small **seed** find energetically favourable positions there, so that they remain attached to the seed. As a result, the motion of atoms towards the seed is not compensated by motion in the opposite direction as occurs in normal diffusion. Thus, in the first place, the seed grows because of the B atoms coming from a halo of the matrix poor in B; secondly, the halo receives B atoms arriving by normal diffusion from regions of the matrix still rich in B. The matrix at the equilibrium concentration develops by surrounding the precipitated grains: this is the equilibrium state (Fig. 5.11).

This process of 'inverse diffusion' causes the alloy to become heterogeneous. However, a detailed analysis shows that, except near the

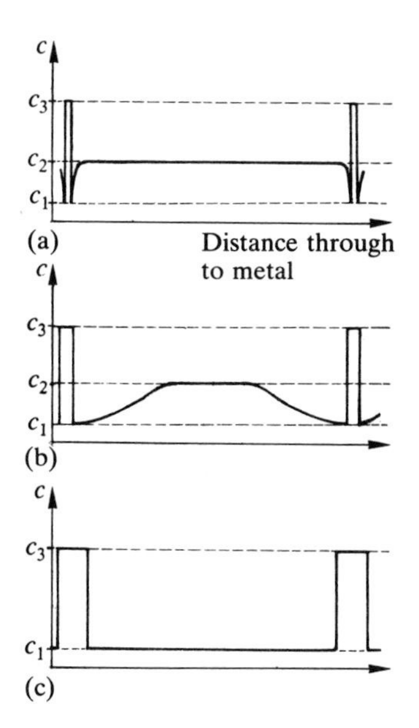

Fig. 5.11. Inverse diffusion. Enlargement of precipitate grains (concentration c_3) in the supersaturated solution (c_2) and the formation of the equilibrium matrix (c_1).

precipitates, which are like potential wells as far as the B atoms are concerned, the diffusion in the space between the precipitates is in fact normal and this region tends to become homogeneous.

Such a mechanism can only operate if the temperature of the metal is high enough to produce an adequate diffusion rate. Consider an aluminium alloy with 5% copper. We have already described (p. 211) the development of the saturated solid solution after a quench. At a low temperature it becomes heterogeneous, but the Al_2Cu precipitate is not formed: the copper atoms gather into very small clusters (10 nm) at the nodes of the solid solution, the **Guinier–Preston** or **GP zones**. The single phase of the solid solution persists, but becomes heterogeneous. This is another inverse diffusion process.

Although the GP zones are very small and numerous, they can only form if the copper atoms travel through distances distinctly greater than those predicted from the normal diffusion coefficient of copper in aluminium at room temperature.

The explanation of this paradox is as follows: the calculated diffusion coefficients are related to the solid solution containing the number of vacancies at thermodynamic equilibrium: at room temperature, this is almost zero, whereas at 550 °C it is about 10^{-4} per atom. Quenching the alloy not only keeps the copper atoms dispersed in the supersaturated solid solution but it also preserves an excess of vacancies which have not had time to disappear towards the surface, the grain boundaries and other sinks. Thanks to the excess vacancies, diffusion in quenched alloys is much faster than it normally is at room temperature. That is why GP zones can be formed: their existence is a good demonstration of the involvement of vacancies in diffusion.

Some applications of diffusion in solids

Diffusion plays an essential role in all solid state reactions and structural transformations, particularly in alloys. As the diffusion coefficient varies very rapidly with temperature, it is easy to understand why the latter must be carefully controlled in all high-quality metallurgical processes.

We mentioned in the last section the precipitation that occurs in supersaturated solid solutions. The reverse effect is the homogenization of alloys after their production by casting the fused metal and before putting them into use. We showed in a special case (Box 21), that the concentration of an element in a heterogeneous metal is a function of the parameter $x/\sqrt{Dt}$. This is a general relationship, easily applied and extremely useful. If, for example, we need to eliminate heterogeneities of dimension x, the treatment times at a given temperature vary as the square of x. In addition, the treatment time is also inversely proportional to the diffusion coefficient: if this doubles for a temperature rise of 10°C, the treatment time is halved.

When two metal components adhere to each other, there is a transition zone between them formed by mutual diffusion. This clearly requires the surfaces (a) to be very clean, (b) to have shapes that fit each other exactly, and (c) to be pressed firmly together. Even if all these conditions are satisfied, however, diffusion will only occur if the metals are mutually soluble. Thus, lead is completely insoluble in iron: an iron sheet quenched in a bath of melted lead emerges from it without retaining a surface layer of lead. This is the opposite of what happens when the iron sheet is plunged into a bath of melted tin or zinc (tinplate, galvanized iron), since tin and zinc are soluble in iron.

The **brazing** of two metal components involves interposing a thin liquid film of a fusible alloy (Pb–Sn or Cu–Ag) between them. The elements in this alloy diffuse into the components, giving the brazed joint a good mechanical strength.

Sintering enables a compact and solid metal component to be obtained from fine powders of one or more metals, and is achieved by compressing a block of the powders while maintaining it at a high temperature. Densification occurs because diffusion takes place between the grains in contact: high temperature treatment homogenizes the product and enables alloys to be obtained that are difficult to produce by solidification from the melt.

Diffusion can also occur in non-metallic solids. The adhesion of the iron in **reinforced concrete** is due to the diffusion of the surface layer of iron oxide both into the iron and into the calcium silicates of the concrete.

Cementation

A mild steel component (i.e. one with a low carbon content) is held at a high temperature for several hours in contact with powdered carbon or in a CO-rich atmosphere. The surface of the iron is covered with the carbon, which then diffuses into the iron, while its surface concentration is maintained constant by the action of the carbon monoxide. The carbonized metal layer gradually increases in thickness at a rate proportional to the square root of the time of treatment, as indicated above. Eventually, when the layer has become about 1 mm thick, the steel is quenched: the carbonized layer is then transformed into martensite (p. 213) and the component thus has a hardened surface. This is **cementation** or **case-hardening**, a technique long used in the steel industry.

The surface of steel can also be hardened by **nitriding**: nitrogen is made to diffuse into the iron, producing grains of extremely hard iron nitride. This process is used for hardening the pinions used in car gearboxes.

The diffusion of carbon or nitrogen into iron are examples of completely asymmetrical pairs. If a carbon block is placed in contact with iron, the carbon diffuses into the iron because an interstitial iron–carbon solid solution can exist (SM, p. 94). The iron, on the other hand, is completely insoluble in the carbon and cannot diffuse into it.

Oxidation of metals

Most metals exposed to the atmosphere become covered with an oxide layer even at room temperature and the process generally occurs more quickly as the temperature is raised. The applications of a metal are partly determined by the nature and degree of the oxidation that takes place, which in turn depend crucially on diffusion. Out of the many oxidation processes that occur, we take a look at one which is quite common.

Consider a metal whose surface is covered by a homogeneous and continuous oxide layer. The metal and the oxygen in the atmosphere are therefore separated from each other by this layer (Fig. 5.12), so that it can only grow if the reacting atoms diffuse through the oxide. At the internal metal–oxide interface, an atom or, more precisely, a metal ion M^+ and an electron, leave the metal. Both diffuse through the oxide. At the external oxide–oxygen interface, the electron transforms an oxygen atom into an O^- ion and the M^+O^- pair is deposited on the oxide crystal and makes it grow. The layer therefore grows on its outer surface.

The rate of growth is determined by the slowest of the complex diffusion processes of the M^+ ion and the electron. Here, this is the diffusion of the

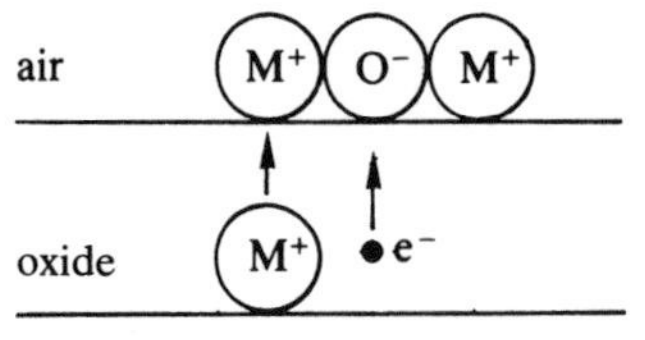

Fig. 5.12. Growth of the oxide layer by the diffusion of metal ions through the layer.

metal ions since the electron is much more mobile. The diffusion is assisted by defects in the oxide crystal, which are vacancies, or missing ions. As a result, diffusion is far from being negligible even at room temperature. The ion flux across the layer is proportional to the concentration gradient of the ions and, since the difference between the concentrations at the two surfaces is constant, the diffusion rate decreases as the layer becomes thicker. The oxidation rate is indeed observed to fall as time goes on: the thickness of the layer increases as the square root of the time.

For some metals, it is not the metal ion but the oxygen ion which diffuses most quickly through the oxide. In this case, the $M^+ + O^-$ reaction takes place at the inner interface: the layer then grows at its base and no longer on its outer surface.

The condition for fast growth of the oxide layer is that the diffusion through the layer should be easy, which implies that many vacancies should be present. This is so in oxides with non-stoichiometric compositions. Thus, iron oxidizes very quickly above 600°C because ferrous oxide FeO, which is only stable above 560°C, is particularly rich in vacancies. Conversely, the oxides Cr_2O_3 and Al_2O_3 contain few vacancies: the layer grows slowly and even this growth almost stops as soon as its thickness exceeds ten nanometres or so. In these cases, an oxide layer protects the metal from further oxidation. This is what happens in the so-called stainless steels, Fe–Ni–Cr: these are covered with a thin film of Cr_2O_3 which completely seals it. In the Ni–Cr superalloys with Al, used because of their mechanical strength at high temperatures (p. 233), the protective layer is of alumina.

Oxidation is only one of a whole family of chemical reactions taking place in the solid state. All of them are controlled by diffusion, the process that enables the reacting atoms to encounter each other. Diffusion in the solid state is a difficult process, however, so that the reactions are relatively slow: they are faster in liquids and faster still in gases.

Chemical reactions and structural transformations in the solid state are manifestations on a macroscopic scale of the perpetual 'jostling' of atoms, which increases as the temperature rises and is in fact a measure of the temperature. Yet instruments of the highest resolving power, electron microscopes or X-ray diffractometers, give us a picture of a perfectly ordered crystalline solid.

The two points of view are not incompatible because by far the majority of the atoms in a crystalline solid are merely vibrating about fixed positions, and it is the order displayed by these that X-ray and electron diffraction are revealing.

Nevertheless, out of all the many millions of vibrations that each atom makes every second, there is one now and again which, because of its large amplitude or because of a fortuitous correlation between the movements of its neighbours, enables it to escape and become lodged in a neighbouring site. Such displacements become more frequent as the mean atomic agitation increases. They are more likely to occur in perturbed regions of the crystal and near the few empty sites persisting in the close-packed structure of the solid.

On a time scale related to the period of atomic vibrations, such displacements are very rare events indeed, yet it is these events that underlie many of the practical properties of solids and make it possible to produce many of the structures most sought after by technologists.

Bibliography

Many of the widely available textbooks on elementary physics are valuable sources of information on the properties of solids, but the following selection is provided for readers requiring more specialized treatments.

Bacon, G. E. (1981). *The architecture of solids*, Wykeham Science Series, Taylor and Francis, London.

Friedel, J. (1964). *Dislocations*, Pergamon Press, Oxford.

Gordon, J. E. (1979). *The new science of strong materials*, Pitman, London.

Kittel, C. (1986). *Introduction to solid state physics*, (6th edn), John Wiley and Sons, New York.

Martin, J. W. (1972). *Strong materials*, Wykeham Science Series, Taylor and Francis, London.

Reif, F. (1965). *Statistical physics*, Berkeley Physics Course vol 5, McGraw-Hill Book Company, New York.

Rosenberg, H. (1988). *The solid state*, (3rd edn), Oxford Physics Series, Oxford University Press.

Wulff, J. (ed.), (1964, 1965). *The structure and properties of materials*, in four volumes: vol. 1, Structure; vol. 2, Thermodynamics of structure; vol. 3, Mechanical behavior; vol. 4, Electronic properties. John Wiley and Sons, New York.

Index

acceptor atoms 106
acoustic branch 36
acoustic waves 39
ageing of steels 217
alloys
  ductility 208
  ferromagnetic 168
  refractory 232
  resistivity 99
alnico 168
α-iron
  carbon diffusion in 189
  in steels 212
aluminium
  Debye temperature 29
  Einstein temperature 29
  elastic properties 182
  heat capacity 7
aluminium alloys 210, 259
aluminium as acceptor 104, 108
aluminium as electrical conductor 99
amorphous semiconductors 123
anelasticity 187
angular momentum 138
  orbital 138
  spin 138
  total 140
anharmonic vibrations 45
anisotropy 23, 58, 60, 177, 186
annealing 213
anomalous dispersion 61
antiferromagnetism 156
Arrhenius's law 102, 249, 253
arsenic as donor 104
ausforming 214
austenite 212
Avogadro's number 3

band
  conduction 105
  forbidden 90
  valence 95, 105
band structure 94, 105
band theory 88
  comparison with classical theory 100
  conductors 95
  insulators 94
  semiconductors 105
barium titanate 66
Barkhausen effect 165
base (transistor) 118
BCS theory of superconductivity 130
beryllium, heat capacity 7
β–alumina 72
binding energy, electron 101, 104
Bitter experiment 161
blue brittleness 217
Bohr magneton 138, 141
Boltzmann factor 145
Boltzmann's constant 2
Bose–Einstein distribution 10, 27
Bose–Einstein statistics 24
Bragg bubble raft 201
Bragg equation 90
Bragg reflection 31, 92
  electrons 89
  X-rays 31
brazing 260
Brillouin function 147, 151
Brillouin zone 39, 91, 93, 95–6
  defined 21–2
  introduced 19
brittle fracture 220
brittleness 176, 190
  blue 217
bulk modulus 180

caesium
  compressibility 45
  thermal expansion 45
carbon
  diffusion in α-iron 189
  see also diamond; graphite
carrier density, semiconductors 102, 110–15
carrier mobility 70, 102, 104
carriers of charge 70
case-hardening 261
Castaing electron microprobe 253
cementation 261
cementite 214
ceramics 234
Charpy V-notch impact test 227
chromic oxide coating 233, 262
chromium 229
chromium dioxide 155
cleavage 220, 222
climb, dislocation 231
cobalt 148
  magnetic properties 155
coefficients of expansion, table 45
coercivity 164
cohesion, intergranular 225, 229

cold-working 186, 206
collector (transistor) 118
Comet aircraft 239
compressibilities, table 45
compressibility 179–80
computers 122
Concorde, alloy for 212
conduction
  band theory 88, 93, 95
  electrical 56, 69
  ionic 71
conduction band 105
conductivity
  electrical 56, 59
  electronic (metals) 73
  thermal 46
  thermal (gases) 51
conductors, organic 134
constantan 99
Cooper pairs 130
cooperative phenomena 152
copper
  band structure 93
  conductivity 102
  Debye temperature 29
  Einstein temperature 29
  elastic properties 182
  electrical conduction 95
  electron drift velocity 74
  heat capacity 7, 12
  non-superconductor 127
  resistivity 56
  unit cell 193
  use as conductor 99
  work-hardening 206
crack initiation 225, 236
cracks, Griffith theory 223
creep 229
creep-resistant materials 232–4
critical exponent (magnetism) 152
critical temperature (superconductors) 127
crystal defects 173
  and electrical conductivity 97–100
  and mechanical properties 188, 199
  and thermal conductivity 49, 54
crystals
  ferroelectric 66
  ionic 42, 56, 61, 72
  piezoelectric 62
  polar 64
  pyroelectric 65
Curie constant 144
Curie point 149, 155
Curie's law 144, 146
Curie–Weiss law 149
damping of oscillations 188
de Broglie relationship 49
Debye frequency 26
Debye model 13, 24–9
Debye temperature 24, 28
  in resistivity 98
  in superconductivity 132
  table of values 29
  in thermal conductivity 54
degrees of freedom 6
demagnetizing field 168
density of electron states 79
  semiconductor 109
diamagnetism 141
diamond
  compressibility 45
  Debye temperature 29
  Einstein temperature 29
  electrical insulator 101
  heat capacity 7, 12
  photoconductivity 103
  thermal conductivity 55
  thermal expansion 45
dielectric breakdown 72
dielectric constant 57
dielectrics 57
  in electric field 59
diffraction
  electron 89
  neutron 31–3, 156
  X-ray 31–3, 211
diffusion coefficient 245
diffusion equation 251
diffusion in liquids and gases 242
diffusion in solids 231, 243–63
  interstitial 189
  inverse 257
dilatometry 45
diode, p–n 115
dipole moment
  electric 58–9
  magnetic 137
dislocation, edge 199
dislocation climb 231
dislocation glide 203
dislocations
  origins 202
  and plastic deformation 199
  sources 203
dispersion 17, 23, 39
  anomalous 61
  phonon 33
dispersion curves 33, 37–41, 51
domain boundaries 161, 165, 170
domains, ferroelectric 67
  ferromagnetic 160

donor atoms 104
doping 103
drain (transistor) 120
Drude–Lorentz theory 74
ductile–brittle transition 227
ductile fracture 226
ductility 176 190
theory 202
ductility in alloys 208
Dulong and Petit's law 3, 6–7, 12, 23, 28–9, 41–2
duralumins 210
dysprosium
ferromagnetic properties 155

edge dislocation 199
effective mass 106
eigenstates, electron 79
Einstein temperature 11
table of values 29
Einstein's model 8, 23, 29
drawbacks of 12
elastic limit 173, 176
calculated 196
experimental 198
table of values 182
elastic waves 13
interaction with electrons 99
in monatomic crystal 15
in polyatomic crystal 39
and thermal conductivity 48
elasticity 173
and atomic structure 183
mechanical theory 177
electrets 66
electric field
action on dielectrics 59
action on solids 56
comparison with magnetic field 136
electric polarization 59
electric susceptibility 59
electrical conductivity 69
electrolytic conduction 71
electron 70
magnetic moments 168
electron diffraction 89
electron in a box 77
electron spin 138
electronic conductivity 73
electronic polarization 61
electrons as carriers 70
emitter (transistor) 118
endurance limit 235
energy
of crystals 3, 23
exchange 153
of free electrons 75, 82
of gases 1, 6
internal 1
of linear oscillator 2, 9
magnetic 143
of three-dimensional oscillator 6, 8
energy gap 90, 92, 102, 113
energy quantization 30
ENIAC 122
equation of continuity 250
equipartition of energy 3
error function 252
exchange energy 153
expansion, thermal 43
extrusions 237

fatigue 234
fatigue fracture 236, 238
Fermi–Dirac distribution 85, 108
Fermi level 82, 109
calculation 83
Fermi sphere 82, 85, 95
Fermi wave vector 82
ferrimagnetism 158
ferrite (α-iron) 213, 214
ferrites 159, 168
ferroelectricity 66
ferromagnetic elements 155
ferromagnetic properties, table 155
ferromagnetism 147
ferroxdur 168
Fick's law 246–8, 253
forbidden band 90, 92
fracture 220
brittle 220
ductile 226
frequency, Debye 26
frustration 157

gadolinium
ferromagnetic properties 155
γ-iron 212
garnets 155
gas constant 3
gases
heat capacity 1
thermal conductivity 51
gate (transistor) 120
gemstones, detection 155
germanium as semiconductor 101–2
carrier density 113
g-factor 140

glasses 123
  fracture 220, 224–5
  toughened 225
glasses as semiconductors 124
glide, dislocation 203
gold, heat capacity 7
GP (Guinier–Preston) zones 210, 259
grain size 214–15
  and plastic deformation 206
graphite 135
grey iron 225
Griffith theory of fracture 223–4
group velocity 18
gyromagnetic ratio 139

hard magnetic materials 167
hardening
  precipitation 209
  structural 211
  work 192, 205–6
harmonic oscillator 2, 6, 44
heat capacity 1
  Debye model 13, 25
  Einstein model 8
  electronic 75, 85
  of gases 1
  molar, defined 1
  of monatomic crystal 3
  of polyatomic crystal 33
  table of values 7
Heisenberg's uncertainty principle 9
helimagnetism 157
holes 107, 112
hysteresis 163

impurities and mechanical strength 216
impurities and resistivity 99
impurity clouds in dislocations 215, 217
inelastic neutron scattering 31–3, 40
infra-red absorption 42
insulators 56
  band theory 94
integrated circuits 120
intercalation compounds 135
interdiffusion 243
internal energy 1
internal friction 188
interstitial diffusion 243
interstitial solid solutions 208
intrusions 237
invar 46
inverse diffusion 257
ionic conduction 71
ionic crystals 42, 56, 61, 72
  brittle fracture 220
  cleavage 220, 222
  high conductivity 72
ionic polarization 61
ions as carriers 71
iron
  diffusion of carbon in 189
  diffusion of nickel in 244–5
  ductility 190
  elastic properties 182
  ferromagnetic properties 155
  ferromagnetism 148
  magnetically soft 166
  Young's modulus 8
iron garnets 159

jet engines 234
junction transistor 118

kinking in zinc 218
Kirkendall effect 257

Landé *g*-factor 140
Langevin function 147
lead
  Debye temperature 29
  Einstein temperature 29
  elastic properties 182
  heat capacity 7
linear chain, vibrations
  diatomic 35
  monatomic 15
lithium
  compressibility 45
  thermal expansion 45
Lüders bands 216–17

magnetic field, action on solids 136
  comparison with electric field 136
magnetic moment
  atomic 137
  current loop 137
  of electron 138
  orbital 138
  spin 138
magnetic permeability 166
magnetic susceptibility 144, 166
magnetite 147, 158
magnetization
  calculation of 144
  spontaneous 149, 150, 160
magneton, Bohr 138, 141
magnets, permanent 137, 167

manganese oxide 156
maraging steels 214
martensite 213
martensitic transformations 212
mean field approximation 148, 153
mean free path
  electron gas 74–5
  molecules in gas 51–4
Meissner effect 129, 130
mercury, superconducting 127
metal fatigue 235
metals
  brittle fracture 225
  conductivity 73, 95
  oxidation 261
mobility
  of carbon atoms in iron 190
  of carriers 70
modes of vibration 13–14
  acoustic 39
  crystal 18, 23, 30
  electron in box 79
  optical 40, 42
  rod 13–14
molecular field, Weiss 148, 153
molybdenum
  Debye temperature 29
  Einstein temperature 29
mumetal 166

necking 191, 226
Néel temperature 156
neutron diffraction 31–3, 156
neutron scattering, inelastic 31–3, 41
nichrome 99
nickel
  band structure 93
  compressibility 45
  diffusion into iron 244–5
  ferromagnetic properties 148, 155
  thermal expansion 45
nitriding 261
non-polar dielectrics 58, 61
normal modes, *see* modes of vibration
n-type semiconductors 103

Ohm's law 70
optical branch 37, 40, 42
orbital angular momentum 138
orbiting electron 138
organic conductors 134
oxidation of metals 261

paramagnetism 142–7
Pauli exclusion principle 82, 91, 140
permalloy 166
permanent magnets 137, 167
permeability, magnetic 166
permittivity, relative 60
permittivity of free space 60
Petch's equation 207
phonon energy 23
phonons 23, 30, 31, 42, 49
  scattering of electron waves 99
photoconductivity 103
photoelectric threshold 102
photoresistors 103
piezoelectricity 62, 187
pinning of dislocations 189
Planck's constant 9
plastic deformation 176, 190
plastics 191
p–n junction 115
p–n–p transistor 118
Poisson's ratio 178
  table of values 182
polar molecules 59, 60
polarization 58
  of dielectrics 59
  electronic 61
  ionic 61
  mechanisms 60
polyacetylene 135
polycrystalline materials 186, 193, 204, 206, 226
polygonization 232
polythiazyl 135
positive hole 107, 112
potassium
  compressibility 45
  heat capacity 87
  thermal expansion 45
precession of atomic moments 143
precipitation hardening 209
preferred orientation 186
ptfe (teflon), electret 66
p-type semiconductors 104
pyroelectricity 66

quanta, energy 9, 23
quantization 9, 23, 30
quantum numbers 9
  magnetic 138
quantum theory
  of heat capacity 9
  of metallic conduction 77, 84

quartz
  piezoelectric effect 62, 187
  resistivity 56
quartz oscillator 64

reciprocal lattice 20, 39, 50
refractive index 57
refractory alloys 232
reinforced concrete 261
relaxation time 187
remanence 164
resistivity 56
  defined 70
  residual 99
  variation with temperature 76, 97, 99
rigidity modulus 176, 179–80, 198
rock salt, cleavage 220, 222
rubber 174

sapphire 55
saturation
  ferromagnetic 164, 167
  magnetic 144, 147
scattering
  inelastic neutron 31–3, 40
  X-ray 31
Schmid's law 199
self-diffusion 243
semiconductors 101
  amorphous 123
  carrier density 110
  doped 103
  extrinsic 105
  impurity 105
  intrinsic 101
  n-type 103
  p-type 104
shear stress 180
silica, vitreous 123
silicon
  amorphous 124–6
  carrier density 113
  conductivity 102
  doped 103
  energy gap 101
  intrinsic semiconductor 101
  perfection of crystal 199
  photoconductor 103
silicon carbide 234
silicon–iron 166
silicon nitride 166
silver
  compressibility 45
  Debye temperature 29
  Einstein temperature 29
  non-superconductor 127
  thermal conductivity 55
  thermal expansion 45
sintering 260
slip 193
slip bands 194
slip planes 194
sodium, conduction 95
sodium chloride
  cleavage 220, 222
  electric susceptibility 60
  heat capacity 34
soft magnetic materials 165
solar cells 126
solid solutons 208
solid state electronics
  compared with vacuum tubes 122
source (transistor) 120
spring, extension of 174
steel, mild
  ageing 217
  elastic properties 173
  tensile curve 215
steels 212
  ferromagnetic 168
  maraging 214
stress 173
substitutional solid solutions 208
superalloys 229, 233
superconductivity 126
susceptibility
  electric 59
  magnetic 144, 166

temperature
  critical (superconductivity) 127
  Debye 24, 28, 29, 54, 98, 132
  effect on conductivity 76
  effect of ductility 190
  effect on electron energy 84
  Einstein 11, 29
  Néel 156
  variation of resistivity 76, 97, 99
tempering 213
tensor 60
thermal conductivity 46, 51
thermal expansion 43
ticonal 168
tin, cry of 219
torsional pendulum 188, 190
toughened glass 225
tourmaline, pyroelectric 66
trace elements
  effect on properties 240

transistor
  field effect 120
  p–n–p 118
TTF–TCNQ 134
tungsten
  compressibility 45
  elastic properties 182
  thermal expansion 45
tungsten filaments 233
twinning 219

ultrasonics 64
Umklapp processes 48

vacancies 72
  and diffusion 243, 255
  and resistivity 100
vacuum tubes (valves) 121
  compared with solid state devices 121
valence band 95, 105
valence electrons 93
vibrations
  atomic 8
  linear chain 15, 35
  rod 13

water, electric polarization 61
wave propagation
  in a linear chain 17
  in polyatomic crystals 39
  in a rod 13
wave vector 14
  elastic waves 48
  electron 88, 95
  phonons 49
Weiss domains 160
Weiss molecular field 148, 153
whiskers 201
Wiedemann–Franz law 76
Wöhler diagram 235
work-hardening 192, 205–6
  and resistivity 100

X-ray diffraction 31–3, 211
X-ray scattering 31

yield drop 215
yield points 215
YIG (yttrium iron garnet) 159
Young's modulus 178, 186, 198
  relation to other moduli 78
  table of values 182

zero-point energy 9
zinc, deformation of 218